STP 1016

Implication of Aggregates in the Design, Construction, and Performance of Flexible Pavements

Hans G. Schreuders and Charles R. Marek, editors

ASTM
1916 Race Street
Philadelphia, PA 19103

Library of Congress Cataloging-in-Publication Data

Implication of aggregates in the design, construction, and performance of flexible pavements / Hans G. Schreuders and Charles R. Marek, editors.
(STP 1016)
"ASTM publication code number (PCN) 04-010160-08"—CIP t.p. verso.
Includes bibliographies and indexes.
ISBN 0-8031-1193-2
1. Pavements, Flexible—Design and construction—Congresses. 2. Pavements, Flexible—Testing—Congresses. 3. Aggregates (Building materials)—Congresses.
I. Schreuders, Hans G. II. Marek, Charles R. III. Series: ASTM special technical publication; 1016.
TE270.I46 1989
625.8—dc19 89-326
CIP

NOTE

The Society is not responsible, as a body,
for the statements and opinions
advanced in this publication.

Peer Review Policy

Each paper published in this volume was evaluated by three peer reviewers. The authors addressed all of the reviewers' comments to the satisfaction of both the technical editor(s) and the ASTM Committee on Publications.

The quality of the papers in this publication reflects not only the obvious efforts of the authors and the technical editor(s), but also the work of these peer reviewers. The ASTM Committee on Publications acknowledges with appreciation their dedication and contribution of time and effort on behalf of ASTM.

Printed in Ann Arbor, MI
May 1989

Foreword

The symposium on Implication of Aggregates in the Design, Construction, and Performance of Flexible Pavements was held in New Orleans, Louisiana, 3 December 1986. ASTM Committee D-4 on Road and Paving Materials sponsored the symposium. Hans G. Schreuders, Westvaco Corporation, Charles R. Marek, Vulcan Materials Company, and Ken R. Wardlaw, National Crush Stone Association, served as symposium cochairmen. H. G. Schreuders and C. R. Marek are editors of this publication.

Contents

Overview

A symposium, entitled "Implication of Aggregates in the Design, Construction, and Performance of Flexible Pavements," was held at the Sheraton New Orleans in New Orleans, Louisiana, in conjunction with the 2–5 December 1986 standards development meetings of the ASTM Committee D-4 on Road and Paving Materials. The focus of this symposium was on the characteristics of aggregates that influence the performance of asphalt paving mixtures. Included were presentations on mixture instability problems (for example, rutting, shoving, bleeding, etc.) which are more prevalent today because of increased axle loads and tire pressures. A total of ten papers were prepared and submitted to the D-4 Papers Review Subcommittee in response to the "call for papers" for this symposium. Five of the papers were selected for presentation at the symposium. All ten papers were considered to contain significant information on the symposium subject and were accepted for publication in this resulting ASTM Special Technical Publication.

Crushed stone and sand and gravel are the two main sources of aggregates used in the construction of flexible pavements. These natural aggregates are widely distributed and exist in a variety of geologic environments. However, although widely distributed, natural aggregates are not universally available for consumptive use. Many areas are devoid of sand and gravel. In some areas potential sources of crushed stone may be covered by great thicknesses of overburden which makes surface mining impractical. In other areas, many aggregates do not meet the physical/chemical property requirements for certain end uses, or they may contain deleterious constituents that react adversely with binding agents used to produce concrete mixtures. Finally, many areas having an abundance of natural aggregate find that the aggregate is not available or accessible because of existing land uses, zoning, or regulations that preclude commercial exploitation of the aggregate. As a consequence, flexible pavements must be designed and constructed with economically available aggregates, possessing quality levels that ensure desired performance of the flexible pavements in which used.

Hot-mix asphalt (HMA) mixtures used in flexible pavements contain approximately 90–95 percent aggregate by weight. Because of this, aggregates are the principal consideration in influencing the properties of HMA mixtures.

The first paper by Abdulshafi and Al-Dhalaan discusses utilization of low quality aggregates in flexible pavements constructed in Saudi Arabia and in the Gulf area. The authors describe use of cement-coated coarse aggregate as one effective solution to rutting, bleeding, and shoving of asphaltic mixtures. Results of laboratory testing are provided to support the conclusions of the authors. The benefits of use of cement-coated aggregate include: increased stability, improved resistance to stripping, increased tensile strength, increased resilient modulus and increased fatigue life.

The paper by Barksdale and others describes the performance of an experimental pavement constructed as a quarry access road in Georgia. The flexible pavement consisted of a triple surface treatment over 18 inches of well-compacted crushed stone base. This thin asphalt surface-thick crushed-stone base pavement performed in a very acceptable manner without major distress when subjected to heavy truck traffic for more than seven years and in excess of 1.4 million equivalent 18 kip single axle loads.

The paper by Brown and others presents data obtained from various studies that show the effect of aggregate grading on the performance of asphalt mixtures. Factors influencing performance of asphalt mixtures that are discussed in the paper include: (1) grading, (2) particle shape, (3) maximum aggregate size, (4) compacted lift thickness, (5) mineral filler content, and (6) aggregate quality. The authors concluded that a well-graded crushed aggregate should be used to provide the highest quality asphalt concrete. In addition, the maximum size of aggregate should be increased to provide higher stability, to improve skid resistance, and to reduce asphalt binder content.

The paper by Dukatz describes one of several recent test roads designed to provide information about the effectiveness of pavements constructed with thin asphalt surfaces over thick aggregate base courses. This test road was a section of SR 1508 in North Carolina constructed in 1985 which also serves as a quarry access road. The section consists of 2 in. of asphalt concrete mix (NC I-2) over 13 in. of well-compacted aggregate base course (ABC). After 18 months of service and achieving 60% of the design traffic, the section is performing as expected with minimal distress. The author states that a back-calculated structural coefficient of 0.20 was appropriate for the crushed stone base used in this test section.

The paper by Dukatz and Phillips describes several problems faced by the hot-mix asphalt engineer in predicting the behavior of asphalt concrete mixtures exposed to moisture. The authors recommend modifications to a test procedure for evaluating moisture susceptibility to permit more accurate interpretation of resulting test data. The authors believe that specimens made for determination of moisture susceptibility (tensile strength before and after conditioning) should be prepared at low (4 to 6%), midpoint (6 to 8%), and high (8 to 10%) air void contents, and that all specimens be used in the analysis procedure. The test results of tensile strength versus air voids for both the conditioned and unconditioned specimens should be plotted separately, and the strength at 7% voids should be obtained by graphical interpolation. The authors also suggest that a minimum conditioned tensile strength at 7% voids in conjunction with a minimum retained tensile splitting ratio (TSR) should be specified.

The paper by Hughes and Maupin addresses the increase in moisture damage that has been experienced in asphalt pavements during the past decade. The authors define moisture damage, and discuss the many factors that may cause the damage. Several methods of evaluating mixtures for moisture susceptibility are reviewed. The need for standardization of a predictive procedure (as also advocated in the paper by Dukatz and Phillips) is emphasized. The authors also discuss several methods for reducing moisture damage potential.

Lee and Al-Dhalaan discuss in detail the pavement distress being experienced in Saudi Arabia as a consequence of changing traffic characteristics due to the crash development programs of the country. A feasibility study was performed to formulate rehabilitation alternatives for pavements experiencing severe rutting problems. Potential solutions include: (1) use of reduced asphalt contents, (2) use of coarser aggregate in the asphalt mixtures, (3) improved quality control, and (4) use of sulfur extended asphalt.

Lundy and others investigated the effects of (1) percent fracture, (2) fines content, and (3) aggregate source on the performance of laboratory prepared asphalt mixtures at a temperature of 10°C (50°F). The repeated load diametral test device was used by the researchers to measure mixture performance. Conclusions drawn by the authors are: (1) an increase in required binder content results from increasing fracture levels, and (2) increase in fines content from 3 to 6% reduces required asphalt content. The authors recommend that additional testing be performed at other temperatures to further quantify the effects of fracture level and fines content on asphalt concrete mixtures.

The paper by Selim and Heidari describes a procedure for measuring the susceptibility of seal coats to stripping. Details of a newly developed laboratory test called the Seal Coat Debonding Test (SDT) are provided. Criteria for evaluating the degree of vulnerability of seal coats to moisture damage are also suggested. The test method presented in the paper is simple and may offer a sound quantitative approach for evaluating potential moisture damage in seal coats, and may assist engineers in selecting the "right" materials for use in seal coat construction.

The last paper, by Tseng and Lytton, presents a method to predict the permanent deformation (rutting) in flexible pavements. The method uses a mechanistic-empirical model of material characterization. Material testing is performed in the laboratory to establish three permanent deformation parameters which represent the curved relationship between permanent strains and number of load cycles. Equations are provided to permit analysis of how the three parameters are affected by (1) material properties, (2) moisture and temperature, and (3) stress state. Permanent deformations calculated from the method are compared with results measured in the field, and "reasonable agreement" is shown.

The ten papers in this publication provide excellent information on aggregate factors that influence the design, construction, and performance of flexible pavements. The information will be of value to all highway/materials engineers working to improve the performance of flexible pavement systems within current economic constraints.

Charles R. Marek
Vulcan Materials Company, Birmingham, AL 35253: symposium cochairman and editor.

Ahmed Abdulshafi[1] *and Mohamad A. Al-Dhalaan*[2]

Utilization of Low-Quality Aggregates in Asphaltic Mixtures in the Eastern Province of Saudi Arabia

REFERENCE: Abdulshafi, A. and Al-Dhalaan, M. A., **"Utilization of Low-Quality Aggregates in Asphaltic Mixtures in the Eastern Province of Saudi Arabia,"** *Implication of Aggregates in the Design, Construction, and Performance of Flexible Pavements, ASTM STP 1016,* H. G. Schreuders and C. R. Marek, Eds., American Society for Testing and Materials, Philadelphia, 1989, pp. 4–18.

ABSTRACT: Utilization of low-quality aggregates in the eastern province of Saudi Arabia as well as in most Gulf areas is becoming a necessity. In this area, the definition of low-quality aggregates is more related to surface chemistry characteristics and salt content than to particle strength, porosity, surface texture, organic and fines or any other deficiency measures. Stripping, ravelling, and rutting are the major distress manifestations observed in asphaltic mixtures. Rutting is related to mix flow. Bleeding and shoving problems are also observed but to a lesser extent. One effective solution to the above is the cement coating of the coarse aggregate, thereby eliminating surface chemistry debonding behavior (stripping) and increasing the overall mixture stiffness. Several laboratory testing methods on aggregates (such as specific gravity, water absorption, abrasion, and soundness) and on asphaltic mixtures (such as stability, modulus of resilience, creep and rutting, fracture toughness, and index of retained strength) have been performed. Laboratory test results on cement-coated and uncoated aggregates as well as on their asphaltic mixtures clearly indicate the improvements obtained by cement coating technique. These improvements include more than 30% increase in Marshall stability with flow remaining within specification range, improved coating and resistance to stripping, increased tensile strength and resilient modulus, more than 90% immersion-tensile strength ratio, and increased fatigue life.

KEY WORDS: low-quality aggregates, cement coating, aggregates upgrading, rutting, asphaltic mixtures, stripping

Aggregate, an important component of asphaltic mixtures, is regularly subjected to degradation actions under load and environmental conditions. The structural integrity of asphaltic mixtures is governed by the aggregate quality which, in turn, is a function of such factors as mineral composition, surface texture and chemistry, type and amount of deleterious content, size and shape of the particles, pore phase characteristics, etc. The physical and chemical breakdown of aggregate due to interaction with water and other aggressive environmental factors has been reported as a potential mechanism contributing to pavement distress [*1*]. Furthermore, a major factor contributing to pavement failures is the mechanical degradation during construction and service life. Because aggregates in the upper layer of the pavement structure are subjected to significant abrasive forces of wheel load traction and severe climatic conditions of freeze and thaw cyclic action, aggregate durability and

[1] Vice president, C.T.L. International, Columbus, OH 43204.
[2] Director general, Materials and Research Department, Ministry of Communications, Riyadh, Saudi Arabia.

soundness assume greater importance. At lower pavement layers, degradation results in increased fines percentage, which may reduce the stability and increase frost susceptibility of those layers.

Aggregate Quality

The definition of aggregate quality and expected performance of locally available materials is becoming a control factor in the design of asphaltic pavements. The determination of acceptable levels of inferior-quality aggregate that can be tolerated at various pavement structural layers depends upon the "use" conditions (that is, whether or not in surface, intermediate, or base course), environmental exposure conditions, and loading type and duration.

Several research studies in the United States have investigated the effect of low-quality materials upon the reliability of performance evaluation; among the most recent is the work conducted at The Ohio State University [*2,3*]. In this work, the engineering properties and durability of various aggregate mixtures were investigated using environmental simulation techniques. The results led to the conclusion that the presence of deleterious materials, either in the fine or coarse aggregates, is equally detrimental to pavement performance. An exception to this conclusion was limestone sand, which has been shown to substantially improve the mixture performance. It has also been concluded that, although low-quality aggregate will result in a lower layer equivalency factor (that is, high pavement-thickness requirement), it is possible to counteract this deficiency by certain modifications, such as using limestone sand as a partial aggregate replacement, adding sulfur to the mixture, etc.

Deleterious aggregate has been defined as follows [*2*]: "A piece of aggregate is to be scraped against the bottom of a stainless steel pan using moderate finger pressure. If it leaves a residue on the pan, it is deemed deleterious; otherwise, it is judged good." In that research study [*2*], the shales were kept separate from other deleterious substances, which were mostly soft sandstones with some lumps of hardened clay and minute traces of soft chert. Based on the above definition, a deleterious material content test was conducted and coarse aggregate classified accordingly.

Deleterious fine aggregate was researched using the application of ultrasonic vibrations under controlled load, frequency, and time duration [*3*]. Test response characteristics of high-quality fine aggregates, including ottawa sand and limestone sand, have been identified and distinguished from low-quality sand. Test response characteristics were established by determining the percentage loss in gradation after exposure to ultrasonic vibrations and sieve analysis. However, more data are necessary before test results can be conclusively interpreted to define the levels of deleterious material content within locally available sand.

Beneficiation Methods

Regardless of whether the amount of deleterious materials found in both coarse and fine aggregates can be known, the beneficiation of such materials is possible. General methods available for the improvement of low-quality materials include blending, impregnation, mechanical processing, or coatings, or a combination of these methods. *Impregnation* of porous aggregates by gaseous or liquid-phase monomeric plastics, chemical treatment, and radiation polymerization has been shown to increase the aggregates' properties, especially those of strength and resistance to freeze-thaw cycling [*4–6*]. *Mechanical processing* by such methods as crushing, washing, heavy media separation, and sieving is routinely used; however, other methods of selective processing by removal of soft and undesirable substances have been used for specific practical applications. *Blending* of low-quality aggregates with high-quality aggregates to upgrade material quality has also been used for specific applications. If, for example, the abrasion and skid resistance need to be enhanced for pavement wearing surface course applications, blending low-quality aggregates with slag or scrap steel can be a promising solution.

Coating of low-quality aggregates can be effected by several means: physical, chemical, thermal, use of admixture, or combined processes. Depending upon the "use condition" of the low-quality aggregates, one of the above-cited methods should prove suitable. *Physical and thermal processes* include moderate heating, 420 to 810°C (800 to 1500°F), or high-temperature heating, 420 to 1100°C (800 to 2000°F), using a fluxing agent. This process produces ceramic surfaces on the aggregate particles [*7,8*]. *Chemical processes* involve coating the particles with thin films of plastics, organic compounds, surface active agents, etc. This process enhances adhesive properties, improves durability, and increases soundness and abrasion resistance of the aggregates.

Admixtures have been used extensively to impart improved properties to low-quality aggregates for use in highways. Among the admixtures commonly used are lime, bitumen, and cement [*9*].

Beneficiation of low-quality aggregates using a cement-coating method is becoming a recognized technique. A laboratory/field study [*10*] has been performed to investigate the potential of cement-coating techniques for mixture performance improvement and practical field applications. In that study, powder cement was added in quantities of 2% to 10% by weight of the aggregates, mix dry, and then water was added to the mixture in a designed water/cement ratio. Mix variables and coating techniques have been optimized in order to achieve the final desired properties and goal of upgrading the aggregate quality, meeting Marshall mix design criteria, enhancing performance behavior and, at the same time, minimizing field application problems such as sticking, coating and bonding deficiencies, segregation, high voids content, etc. This study concluded that cement coating is a viable method that will improve asphaltic mixture properties and minimize potential failure from fatigue cracking and rutting. Because of cement-coating efficiency in upgrading low-quality aggregate properties, the authors believe that this method is more economically and practically feasible for application to asphaltic mixtures than the other methods mentioned above.

Problem Statement for the Locally Available Aggregates

The Kingdom of Saudi Arabia has built over 35 000 km of roads during the past two decades. At present, highways, especially in the eastern province, are showing different degrees of pavement distress: severe rutting, stripping and raveling problems, and to a lesser degree, fatigue cracking. Among the factors contributing to these distresses are the ever-increasing traffic axle loads, high prevailing summer temperatures, and the hydrophilic characteristics of the local aggregates. Other factors include the quality of the produced aggregates as a result of quarry operations of intermixing different quality aggregates, crushing and sorting techniques, and deleterious material content. The specification for the quality of aggregates in the Kingdom of Saudi Arabia is set by specifying approved material sources and passing other American Association of State Highway and Transportation Officials (AASHTO) or American Society for Testing and Materials (ASTM) testing requirements as requested by the site engineers as follows:

Loss of Sodium Sulfate Soundness Test (AASHTO T104)	10% maximum
Loss of Magnesium Sulfate Soundness Test (AASHTO T104)	12% maximum

Loss by Abrasion Test (AASHTO T96)	40% maximum
Thin and Elongated Pieces, by Weight (larger than 1 in., thickness less than 1/5 length)	5% maximum
Friable Particles (AASHTO T112)	0.25% maximum
Sand Equivalent (AASHTO T176)	45% minimum

If the above specificiation is not met, the aggregate is judged to be of low quality. However, depletion of quality aggregate resources and the high cost of its transportation to road construction areas have caused the quarry operators to mix high- and low-quality aggregates. Based on sampling and testing aggregates from seven quarries in the eastern province (shown in Fig. 1), the majority of tests indicate high contents of deleterious material, high water absorption values, and low abrasion resistance. The solution to the above problem includes exploration for natural sources of aggregates within the large landmass of the kingdom, improving crushing and sorting techniques, and upgrading the aggregate quality by using additives such as lime or cement. The latter is the topic of this paper.

Aggregates Location

Geological studies of the eastern coastal regions indicate that bedrock formation has resulted in a varying degree of quality aggregates [*11,12*]. Data on physical and mechanical properties of rocks from the Jubail, Dhahran, and Hofuf areas are presented in Tables 1 and 2 [*13*]. Major categories of petrological classification of rock types are [*12*]: fine-grained crystalline limestone (micrite); medium-grained crystalline limestone (sparite); and detrital limestone (biosparite).

Rock mass properties were difficult to assess in detail due to the limited amount of exposure arising from surface ripping techniques. However, observations noted in several quarries make an appraisal of the general conditions possible [*12*]. Bedding was apparent on several occasions, and individual units ranged from 0.25 to 1.5 m in thickness. Where larger working surfaces were visible, inter-unit clay bands were in evidence; these were usually about 20 cm in thickness and composed of light- to dark-brown soft plastic clay. The bedding, when observable, was nearly always horizontal except at one location where a slight dip of 2 to 3 deg was discernible to the naked eye.

No definite or dominant joint patterns were in evidence. However, rock units near the surface, when relatively freshly exposed, did show a "blocky" (equidimensional) nature on

TABLE 1—*Physical and mechanical properties of source rocks, Jubail, Dhahran, and Hofuf quarries.*

Location	Material	Specific Gravity	Absorption	Los Angeles Abrasion	Soundness
Jubail	sand (SP)	2.64–2.71	0.9–2.4	14–40	7.1–23.4
Dhahran	limestone aggregates	2.59–2.71	...	...	...
Hofuf	limestone aggregates	...	3.9–5.6	31–44	...

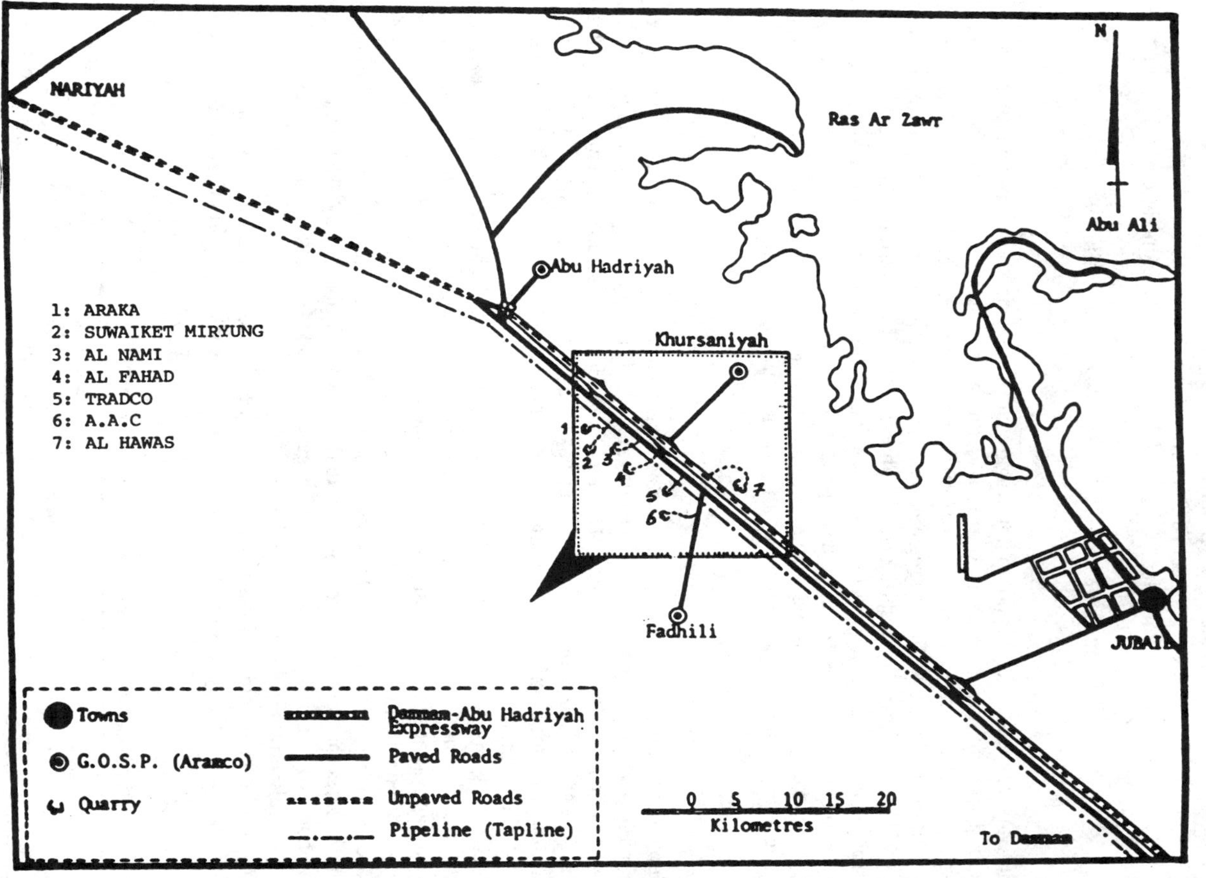

FIG. 1—*General location of study region: Al Hoty-Stanger, Ltd.* [11].

TABLE 2—*Physical and mechanical properties of source rocks, Jubail area quarries detailed test results.*[a]

Location	Bulk Specific Gravity, SSD[b]	Water Absorption, %	Slake Durability Index, %	Point Load Test Strength Index,[c] MPa				Sulfate Soundess, %	
				A	B	C	D	Sodium	Manganese
Al Raka	2.57	1.3	99.1	10.0	9.5	18.5	...	0.2	0.2
	2.54	1.5	99.2	11.5	10.0	20.6	...	0.3	0.3
Suwaiket	2.46	2.0	98.8	12.5	11.7	10.5	...	0.2	0.3
Miryung	2.56	1.0	99.3	13.2	10.0	13.8	12.5	0.3	0.3
Al Naimi	2.42	2.4	99.2	8.8	6.1	7.9	...	0.2	0.2
	2.58	1.1	99.1	12.8	16.4	15.5	...	0.2	0.1
Al Fahad	2.58	0.9	99.1	12.4	12.4	17.2	...	0.1	0.2
	2.47	2.2	99.1	9.7	10.3	11.0	...	0.1	0.1
Tradco	2.57	1.7	98.0	14.5	8.8	10.7	13.1	0.2	0.3
A.A.C.	2.48	2.1	99.0	13.0	8.8	11.7	...	0.8	0.5
Al Hawas	2.56	1.4	99.1	13.1	18.7	15.1	11.7	0.1	0.3

[a] All tests performed according to relevant AASHTO standards.
[b] SSD = saturated surface dry basis.
[c] A, B, C, D = tests on four representative samples.

occasion. Whether this fracture pattern was directly related to planes of weakness within the rock mass was not clearly evident. (Fractures tended to be less than 200 mm in width.) Many of the fractures displayed a light-brown, friable "sandy" infill, probably of secondary origin, which was moderately weathered. No joint or fracture patterns appeared to be present below these uppermost units.

Voids and cavities were often evidenced and in some cases were quite extensive. Estimated rock strength ranged from moderately strong at uppermost layers to strong in deeper layers.

Regarding weathering, most ripped rocks appeared to be fresh, except for the uppermost unit which was exposed on overburden removal and showed slight to moderate weathering. (A slight brownish discoloration often penetrated to a depth of a few centimetres.) Stockpiled raw feed was also seen to have undergone similar slight weathering/alteration at several sites.

Aggregate, as was to be expected, reflected the properties of the source rock; particle shape was found to be sub-angular to angular. Surface texture ranged from rough and crystalline to relatively coarse. Stockpiled material, however, often showed slight weathering where it had been accumulated for some considerable time. This took the form of a brownish discoloration and surface pitting.

Physical and Mechanical Properties

Aggregate samples from five of the quarries shown in Fig. 1 were obtained and tested in the laboratory. Tables 3 and 4 present a summary of the results of these tests.

Based on site visitations and test results shown in Tables 2 and 3, aggregates from three quarries were identified as being of low quality, medium quality, and high quality. This is required in order for us to study the characteristics of aggregates in each category in view of evaluating the effectiveness of the improvement techniques. However, it should be noted that delineation of those categories is primarily based upon aggregate source, chemical, physical, and mechanical properties rather than mixture characteristics and performance behavior. For example, although the aggregates may be classified as good quality, a stripping (debonding) problem could occur when used in an asphaltic mixture.

TABLE 3—*Summary of aggregates' physical mechanical properties test results.*

		Sieve Size, mm							Guideline Specifications (for F)
		40	25	20	15	12.5	10	F	
Percent passing	$\bar{x}$[a]	0.75	1.15	0.92	1.87	1.4	1.35	14.18	0.5% max.
	n[b]	4	2	6	3	2	2	7	(2%)
	s[c]	0.25	0.5	0.3	1.13	0.28	0.9	6.1	
Clay lump	$\bar{x}$	0.2	0.45	0.3	0.5	0.5	0.73	1.2	0.25% max
	n	4	2	6	3	2	3	7	
	s	0.14	0.35	0.2	0.26	0.14	0.15	0.77	(NA)
Flakiness Index	$\bar{x}$	22.3	14.5	19.2	26.67	32	35.67	...	5% max.
	n	4	2	6	3	2	3	...	
	s	12	3.5	5.4	5.0	9.9	6.51	...	(NA)
Chloride Content, %	$\bar{x}$	0.01	0.01	0.012	0.01	0.015	0.01	0.0125	0.03 max
	n	4	2	6	3	2	3	7	
	s	0.0	0.0	0.004	0.0	0.007	0.0	0.0046	(0.05)
Sulfates Content, %	$\bar{x}$	0.24	0.185	0.115	0.17	0.095	0.073	0.195	0.5 max
	n	4	2	6	3	2	3	7	(0.04)
	s	0.14	0.021	0.074	0.17	0.05	0.055	0.119	
Bulk Specific Gravity,	$\bar{x}$	2.53	2.69	2.53	2.5	2.58	2.57	2.6	...
	n	4	2	6	3	2	3	7	(...)
	s	0.064	0.23	0.064	0.084	0.035	0.044	0.022	
Water Absorption, %	$\bar{x}$	2.15	2.1	2.12	3.13	2.7	2.7	2.08	2% max.
	n	4	2	6	3	2	3	7	
	s	0.53	0.42	0.5	1.36	0.14	0.7	0.41	(1%)
Fines Content, %	$\bar{x}$	0.475	0.5	0.33	1.3	0.8	0.97	8.67	...
	n	4	2	6	3	2	3	7	
	s	0.36	0.14	0.1	1.23	0.14	0.57	3.45	(...)
Slake Durability, %	$\bar{x}$	98.25	98.6	98.5	96.3	97	97.1	...	99% min
	n	4	2	6	3	2	3	...	
	s	0.94	0.42	0.35	0.58	0.0	0.75	...	(...)
Elongation Index, %	$\bar{x}$	34	24	29.8	31	26	34	...	5% max
	n	4	2	6	3	2	3	...	
	s	3.74	8.5	4.88	7.2	21.2	19.16	...	(NA)

[a]$\bar{x}$ = samples average (seven quarries).
[b]n = number of quarries.
[c]s = standard deviation.

TABLE 4—*Summary of aggregate test results (for asphalt concrete).*[a]

Location	Los Angeles Abrasion Test						Sulfate Soundness Test, %			
				100/500 RETD[b]			Sodium		Manganese	
	Grading A	Grading B	Grading C	A	B	C	Coarse	Fine	Coarse	Fine
Al Raka	...	41	40	...	0.27	0.3	1.8	7.0	2.2	7.4
Suwaiket Miryung	37	38	36	0.3	0.29	0.28	1.5	7.8	1.1	9.7
Al Naimi	34	30	...	0.26	0.27	...	1.2	3.4	1.3	4.0
Al Fahad	35	36	41	0.29	0.28	0.29	0.9	3.0	0.8	3.0
Tradco	...	39	42	...	0.26	0.26	7.1	8.3	8.9	8.4
A.A.C.	...	32	34	...	0.25	0.24	2.3	7.0	1.6	6.9
Al Hawas	32	33	39	0.28	0.27	0.33	1.5	1.6	0.6	1.0
Guideline Specifications	40% max			0.25% max			10% max		12% max	

[a]All tests performed according to relevant AASHTO standards.
[b]Ratio of loss. Refer to AASHTO T96 (Note 6).

Laboratory Testing Program

AC Mixture Characterization

Marshall Design Criteria—Table 5 presents laboratory test data on representative samples of low, medium, and high quality aggregates. Asphaltic mixtures were prepared in accordance with the Marshall method, and the test results are given in Table 6. Table 6 shows that the low-quality aggregates (LQA) mixture satisfies the Saudi specification limits; that is, if LQA is not rejected as a material, it would not be rejected in asphaltic mixtures. The Saudi asphalt concrete (AC) mixture limits (more stringent than Marshall criteria requirements, see Table 8) should not be the only criteria used in the quality control schemes.

Short-Term Characterization—The short-term characterization of asphaltic mixtures deals with properties of strength, elasticity, viscoelasticity, and fracture and water susceptibility. Laboratory tests conducted to reflect these properties include indirect tension tests (IT), modulus of resilience (MR), creep compliance ($J(t)$), fracture toughness (K_{Ic}), and stripping (or, alternatively, index of retained strength) tests. These tests have been conducted and for comparison purposes are shown and discussed in other sections of this paper.

Long-Term Characterization—The long-term characterization of asphaltic mixtures deals with its performance in relation to the occurrence of distresses; the most common in Saudi are rutting, fatigue cracking, bleeding, stripping, and ravelling. Laboratory tests conducted to simulate (and quantify) these distresses include fatigue on disks, index of retained strength for stripping, and incremental static-dynamic series for rutting. Bleeding and raveling are controlled by mix proportioning in the Marshall design method. Since stripping test results indicated the moisture susceptibility of LQA mixtures, it was decided to improve aggregates quality and determine the effectiveness of the techniques rather than carry the fatigue and rutting testing program. Test results on LQA asphaltic mixtures are also cited and discussed for comparison purposes in other sections of this paper.

Laboratory Coating Method for LQA

Test data and other research studies cited show that the problem with LQA for use in asphaltic mixtures is related more to surface debonding than to any other feature. The addition of lime or cement to the asphaltic mixture has slightly lessened the stripping problem [*13*]. Cement or lime coating-techniques have been effective in providing the separating surface that bonds equally well to both the aggregates and bitumen. The following laboratory procedure was followed for coating:

TABLE 5—*Laboratory test data of coarse aggregates.*[a]

Aggregate Type	Bulk Specific Gravity, gm/cm³	Apparent Specific Gravity, gm/cm³	Absorption, %	Deleterious Content, %	Abrasion, %	Stripping, %	Sodium Sulfate Soundness, %
Low quality aggregate	2.483	2.619	2.30	11.60	45.6	<95	6.3
Intermediate quality aggregate	2.573	2.642	1.31	2.67	33.3	>95	3.1
High-quality aggregate	2.619	2.694	1.06	1.10	26.6	>95	1.1

[a]All tests performed according to relevant ASTM standards.

TABLE 6—*Test results summary on treated and untreated aggregates AC mixtures.*

Marshall Data	LQA,[a] Untreated	LQA, 3% Cement	LQA, 6% Cement	LQA, 3% Lime	LQA, 6% Lime	Intermediate-Quality Aggregate	High-Quality Aggregate	Bituminous Wearing Course Class "B" Limit
Stability, kg	1500	1950	2100	1200	1170	2100	2010	700 min
Slow, mm	3	3	2.4	4	1.4	2.8	3.8	2.4 to 4.0
Air voids, %	4.5	3.9	4.9	4.3	5.8	4.6	4.5	3 to 5
VMA, %	16	16.8	16.4	17.2	17.3	16	16	...
Density, gm/cc	2.324	2.305	2.307	2.290	2.270	2.241	2.301	...
Asphalt cement, % total mix	5.4	5.3	5.2	5.8	5.4	5.2	5.5	4 to 7
MR data (MPa) × 10^3								
at 22°C	6.473	7.647	8.3	...	...	...	...	...
at 37°C	1.40	3.345	4.18	...	...	...	...	...
Compressive strength kg/cm^2	47.1	67.3	69.6	53.1	52.8	...	...	...
Index of retained strength,[b] 7%	58	91	92	57	50	...	...	...
Stripping potential[c]	95%	no stripping	no stripping	...	...	...	...	95%
Indirect tensile strength (kg/cm^2) dry/wet conditions								
at 1 day	6.5/0[d]	9.3/8.8	9.3/8.9	...	...	...	...	...
at 7 days	9.4/0[d]	10.8/8.9	10.6/9.2	...	...	...	...	...
at 14 days	10.7/3.6	12.6/13.3	13/12.7					

[a]LQA = low-quality aggregates.
[b]ASTM D 1075.
[c]ASTM D 1664.
[d]0-mix deteriorated.

1. Aggregates were dried in an oven according to the ASTM Test Method for Specific Gravity and Absorption of Coarse Aggregate (C 127-84) specifications to a constant weight at a temperature of 110 ± 5°C.
2. Specific gravity and absorption were determined according to ASTM C 127-84 specifications.
3. The aggregates were separated into individual fractions; namely, passing 19-mm sieve but retained on 12.5-mm sieve, passing 12.5-mm sieve but retained on 9.5-mm sieve, and passing 9.5-mm sieve but retained on 4.75-mm sieve.
4. Water required for the coating process was determined based on the amount needed for absorption by the aggregate (percent) and the amount needed for coating the aggregate with either cement or lime. Water for absorption was added based on laboratory tests for absorption of aggregates. A water/cement ratio of 0.2 was found to be adequate for determining the water requirement for coating and hydration of both cement and lime mixes.
5. Each fraction of the aggregate was first mixed with the water required for absorption and thereafter mixed with water and cement or water and lime for a period of three minutes. (Water content was based on water/cement or water/lime ratio.) Mixing time could be reduced to a minute if a mechanical mixing device is used.
6. The mixed aggregates were covered with a plastic sheet and allowed to cure for 48 h. Curing water was sprinkled over the coated aggregates 24 h after coating.
7. Once the aggregate fractions are coated and left for two days (minimum time for hydration), they are considered new aggregates; after coating, a new gradation is considered necessary. Tests carried out in the laboratory did not result in any change in gradation after coating; this lack of response may have been caused by the fact that only small quantities of aggregates were coated at a time. It may also be attributed, in part, to the coating of individual fractions separately. During routine construction works, coating of individual size fractions is neither practicable nor economical. So, when cement-coated or lime-coated aggregates are produced on a mass scale, adjustments may have to be made for the overall gradation of the mix.

Test Results and Analysis

A summary of test results on asphaltic mixtures with coated and uncoated low quality aggregates is presented in Table 6. Several observations can be made from it:

1. The Marshall criteria specification for asphaltic mixtures (stability, flow, percent air voids, percent voids in mineral aggregate [VMA], and asphalt content) could be met regardless of the aggregates quality. However, the index of retained strength indicates that untreated LQA mixtures and lime-treated LQA are more susceptible to water damage than cement-treated LQA. The same also applies when considering stripping potential test results. This appears to substantiate the statement that aggregates quality in the kingdom should be judged in terms of particle surface characteristics rather than strength and density factors. The compressive strength data indicate that there is a slight benefit in using lime or cement for coating aggregates; however, the index of retained strength is the main test to judge LQA improvement. The indirect tensile strength test results also substantiate the statement that cement coating will improve surface characteristics of LQA substantially.
2. The MR of LQA asphaltic mixtures at 22°C increases slightly by using cement coating as a treatment over the untreated mixtures, but for all practical reasons it could be stated that cement treatments will not affect the MR values at 22°C. Although MR values decreased for all mixtures at high temperature (37°C), the cement-coated LQA mixtures doubled that of the untreated mixtures. This is an important result considering the high temperature prevailing in the Kingdom of Saudi Arabia during the summer.

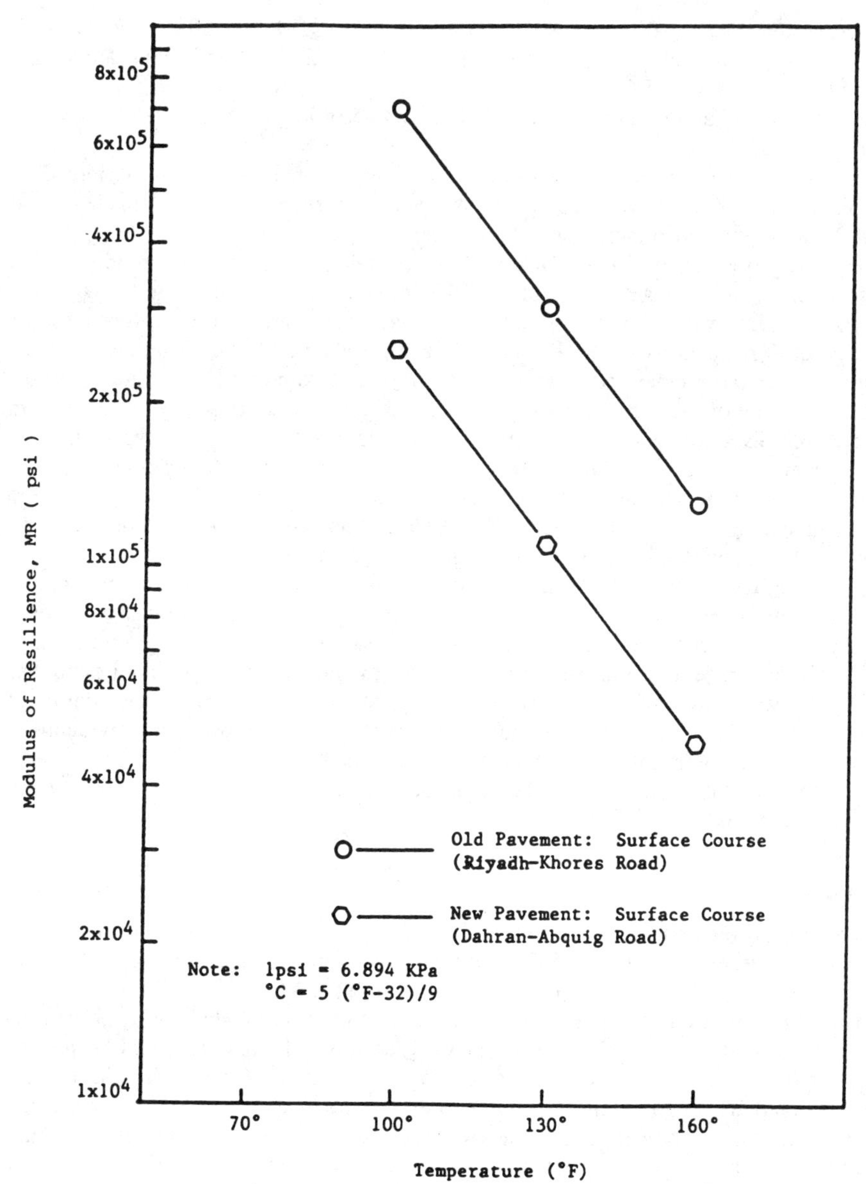

FIG. 2—*MR-temperature curve of extracted cores (high-quality aggregates).*

To better understand the effect of high temperature on MR values, a number of cores were extracted from Riyadh-Khores-Abqaiq roads and tested for MR at high temperatures. The relationship between MR and temperature is shown in Fig. 2. It was concluded from the test results that MR values were lower than expected not only because the asphaltic mixture becomes softer but also because the aggregates in contact with linear variable differential transducers (LVDT) tend to rotate; hence, an additional strain (horizontal component of the rotational vector) becomes part of the total measured strain. Although this factor was observed, it was difficult to assess within the scope of work under the testing program. Consequently, the MR values obtained for the cement-coating treatment at high temperature is considered conservative. In 1985, excessive rutting was observed in the truck lane of the Dhahran-Abqaiq road (built with LQA asphaltic mixture) with the formation of channels up to 15 cm deep. The pavement condition rating is shown in Tables 7 and 8. In 1984, the VESYS III program was used to predict rut depth on the same road given the input information shown in Table 9. Figure 3 presents rut depth prediction that in a one-year period (that is, 1985), there will be as much as a 10-cm rut depth for 20-ton axle loads and 5.5 cm for 9-ton axle loads. Truck traffic count was 2304 per 8-h working period with 54% to 46% split between south- and northbound lanes, respectively. The southbound traffic consisted of loaded trucks of which about 35% were either heavily loaded or overloaded (between 13 and 20 tons).

It is worth mentioning that rutting prediction by VESYS program was governed by the reduced MR, GNU, and ALPHA values at the higher temperatures and axle loading. The excessive rutting measured in the field may indicate that these factors are more severe when occurring in the field than when simulated in the laboratory. However, the main point is that the area's prevailing, high summer temperatures and the high traffic axle loading conditions will result in excessive rutting in a short period of time, as evidenced by both

TABLE 7—*PCR[a] for Dhahran-Abqaiq Road (southbound).*

Section, km	Structural Deduct	PCR	Remarks
0 + 000			start-interchange
	10.50	86.25	Abqaiq-Dhahran
4 + 750			
	10.50	82.50	excessive rutting
6 + 200			overpass
	15.00	84.40	excessive rutting
14 + 200			
	15.00	82.50	excessive rutting
17 + 200			
	10.50	86.75	
18 + 200			
	15.00	83.00	
25 + 700			gas station
	15.00	82.50	
32 + 200			
	15.00	82.50	excessive rutting
40 + 700			gas station
	15.00	82.50	
42 + 900			
	10.50	85.75	
52 + 200			
	10.50	86.25	
57 + 200			
	10.50	86.25	
60 + 200			
	15.00	85.00	excessive rutting
61 + 200			
	4.50	94.50	
64 + 700			end—Abqaiq exit

Average pavement condition rating for the complete section, PCR = 84.75.

[a]PCR = pavement condition rating.

TABLE 8—*PCR*[a] *for Abqaiq-Dhahran Road (northbound).*

Section, km	Structural Deduct	PCR	Remarks
0 + 000			start Abqaiq exit
	4.50	95.50	
4 + 000			
	10.50	89.50	
8 + 500			
	15.00	85.00	
17 + 000			
	15.00	85.00	
32 + 200			intersection
	10.50	89.50	
38 + 000			gas station
	10.50	86.75	
48 + 700			
	10.50	86.75	
58 + 700			distance sign: "9 km to Dhahran"
	10.50	86.75	
60 + 700			
	15.00	82.50	
64 + 700			end
Average pavement condition rating for the complete section, PCR = 86.82.			

[a]PCR = pavement condition rating.

the laboratory testing and field measurements. Finding a solution to the rutting problem is not as easy as it appears to be. It is known that the thicker the pavement, the greater the occurrence of rutting; the stiffer the mix, the more susceptible it will be to fatigue cracking. Design of pavements in this area should consider both fatigue and rutting distresses under prevailing conditions of high summer temperatures, high axle loading, and tire pressure. The current design practice does not consider performance or material or mixtures upgrading.

A laboratory evaluation of the long-term performance of asphaltic mixture with and without cement-coated LQA was conducted in the Kuwait study [*10*]. That study showed that cement coating of LQA enhances performance of asphaltic mixtures in regard to both rutting and fatigue distresses. Those conclusions are consistent with the findings of this research study.

TABLE 9—*Input values for rut depth prediction using Vesys III program (Dhahran-Abqaiq road).*[a]

	Wearing Course Layer		AC Stabilized Base Course Layer		Subgrade Layer
	Test Temperature, °C (°F)				
Parameter	38°C(100°F)	71°C(160°F)	38°C (100°F)	71°C (160°F)	N/A
MR × 10^5, psi	6.6	2.5	4.2	0.77	0.2
Alpha	0.80	0.4	0.6	0.76	N/A
GNU	0.83	0.18	0.062	0.06	N/A

[a]Axle loads considered are: 9, 13, and 20 tons. Structural composition: 5-cm AC wearing course; 20 cm AC stabilized base course.

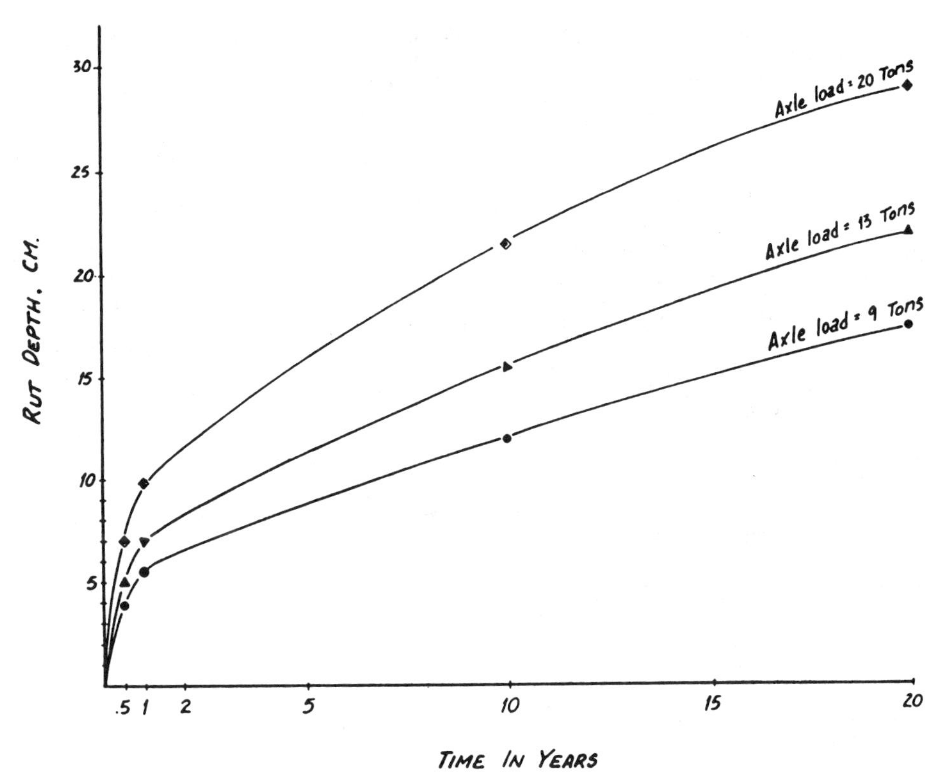

FIG. 3—*Vesys III: Rut depth prediction high-quality aggregates AC mixtures at 38°C (100°F).*

Conclusions

1. Low-quality aggregates in the eastern province of Saudi Arabia are defined in terms of particle surface characteristics. In this regard, meeting Marshall method criteria cannot be used as an acceptance measure for the aggregates quality.

2. There is a need to develop a rational test method for defining levels of aggregate quality.

3. Cement coating of the coarse fraction of LQA will result in upgrading this aggregate for use in asphaltic mixtures. It will enhance short-term and, as indicated in some other research [*10*], it will enhance the long-term characteristics of these mixtures.

4. Cement coating of LQA will require the development of new sets of specification for quality-control measures of asphaltic mixtures. This is necessary since the addition of cement will change requirements for aggregates gradation, percent air voids, and mix density.

References

[*1*] West, T. R., Johnson, R. B., and Smith, N. M., "Tests for Evaluating Degradation of Base Course Aggregates," National Cooperative Highway Research Program Report 98, 1968.

[*2*] Majidzadeh, K. and Ilves, G., "Aggregate Quality as Related to Flexible Pavement Performance," Final Report, EES 512, The Ohio State University Engineering Experiment Station, 1976.

[*3*] Majidzadeh, K., Hou, T. H., and Abdulshafi, A., "Evaluation of Asphaltic Mixtures with Deleterious Fine Aggregates," Final Report, EES 584, The Ohio State University Engineering Experiment Station, 1981.

[4] Manowitz, B., Steinberg, M., Kuback, L., and Colombo, P., "Development of Concrete-Polymer Materials," BNL-13732, Brookhaven National Laboratory, New York, 1969.

[5] Cady, P. D., Blankenhorn, P. R., and Kline, D. E., "Upgrading of Low Quality Aggregates for PCC and Bituminous Pavements," National Cooperative Highway Research Program Report 207, 1977.

[6] Dutt, R. N. and Lee, D., "Upgrading Absorptive Aggregate by Chemical Treatments," *Highway Research Board Record,* No. 353, 1971.

[7] Jewett, C. L., "Colored Granulated Materials," Patent No. 2 379 358, Minnesota Mining and Manufacturing Company, Minneapolis.

[8] Jewett, C. L., "Silicate Cement, Particularly Useful as a Coating," Patent No. 2 378 927, Minnesota Mining and Manufacturing Company.

[9] Majidzadeh, K. and Bovold, F. N., "State of the Art: Effect of Water on Bitumen-Aggregate Mixtures," Highway Research Board, Special Report No. 98, 1968.

[10] Guirguis, H. R., "Cement Utilization in Asphaltic Concrete Paving Mixes in Kuwait," Road Research Center, Ministry of Public Works Department, Kuwait, 1980.

[11] Powers, R. W., Ramirez, L. F., Redmond, C. D., and Elberg, E. L., "Geology of the Arabian Peninsula," *Sedimentary Geology of Saudi Arabia,* U.S. Geological Survey Professional Paper 560-D, 1966, pp. 1–147.

[12] *Aggregate Production Survey,* Abu Hadriyah Region Quarries, Al Hoty-Stanger Ltd., Dehran, 1983.

[13] "Synthesis on Low Quality Aggregates Improvement Techniques," Subtask A-2, Final Report, Resource International, Inc., Weskrville, Oct. 1986.

Richard D. Barksdale,[1] *R. L. Greene,*[2] *A. J. Bush,*[3] *and Charles A. Machemehl, Jr.*[4]

Performance of a Thin-Surfaced, Crushed-Stone Base Pavement

REFERENCE: Barksdale, R. D., Greene, R. L., Bush, A. J., and Machemehl, C. A., Jr., **"Performance of a Thin-Surfaced, Crushed-Stone Base Pavement,"** *Implication of Aggregates in the Design, Construction, and Performance of Flexible Pavements, ASTM STP 1016,* H. G. Schreuders and C. R. Marek, Eds., American Society for Testing and Materials, Philadelphia, 1989, pp. 19–33.

ABSTRACT: An experimental pavement approximately 200 m (650 ft) long was constructed in May of 1978 consisting of a 19-mm (0.75-in.) triple surface treatment over a 460-mm (18 in.) crushed-stone base. By 1985 the pavement had been subjected to about 1.4 million equivalent 80 kN (18 kip) American Association of State Highway and Transportation Officials (AASHTO) single-axle loads neglecting front axle effects. The straight segment of the pavement had a present serviceability rating of 3.0 to 3.5 in 1985. About 6.5% surface cracking had developed mostly within the last two years (1984 to 1985). The pavement is constructed on a clayey and sandy silt subgrade which is relatively dry and has a field California bearing ratio (CBR) of 10. The crushed stone base is well graded, has a maximum aggregate size of 30 mm (1.5 in.), and contains less than 2.0% fines. Field tests performed on the pavement included static plate load tests, Benkelman beam, surface strain measurements, and falling weight deflectometer (FWD) tests.

Using the AASHTO design method, a base course layer coefficient for the upper 305 mm (12 in.) was backcalculated to be 0.18, and 0.16 for the remaining stone base. The elastic modulus of the base was backfigured from the FWD test results to be 406 MN/m^2 (59 ksi) using the Waterways Experiment Station (WES) approach; other methods of backcalculation gave widely varying results. The elastic moduli of the base and subgrade depend upon the test method and also the mean stress and shearing strain to which the material is subjected. Because of the difference in loading conditions, the elastic modulus of the base will be somewhat lower under a dual-wheel, 80-kN (18-kip) axle loading than under the single, 40-kN (9-kip) FWD test loading.

KEY WORDS: field investigations, pavements, pavement design, pavement condition, performance evaluation, performance

Stone has been used throughout history in the construction of roadways. Short portions of the Appian Way, which were constructed with carefully cut stone about 2000 years ago, are still in use. Today, crushed stone is widely used in both unsurfaced and surface treated roads, and also for nonstabilized and stabilized base courses for high traffic-volume pavements. Crushed stone provides the roadway with structural integrity, and in many cases drainage, while minimizing construction costs. However, crushed stone at the present time

[1] Georgia Institute of Technology, Atlanta, GA 30332.
[2] Subsurface Investigations, Inc., P.O. Box 241204, Charlotte, NC 28224-1204.
[3] Waterways Experiment Station, P.O. Box 631, Vicksburg, MS 39180.
[4] Vulcan Materials Company, P.O. Box 7497, Birmingham, AL 35253.

is frequently not used to its full economic advantage in pavement construction because of a general lack of understanding of its behavior characteristics.

A 200-m-long (650-ft) experimental section of pavement was constructed in May 1978 at the Vulcan Materials Co. quarry in Stockbridge, Georgia. This pavement was constructed using a thin, triple-surface treatment over a thick, crushed-stone base. This paper reports the condition and strength of the pavement determined by use of *in situ* field evaluation techniques in 1985 after eight years of heavy service.

Pavement Construction

The top of the subgrade of the pavement was initially strengthened by stabilization with about 160 kg/m^2 (300 lb/yd^2) of coarse No. 57 crushed granite graded aggregate base determined in accordance with the ASTM Specification for Standard Sizes of Coarse Aggregate for Highway Construction (ASTM D 448-80). This stone was added to the top of the subgrade but not mixed in. Because of the relatively high strength of the subgrade, the stone did not penetrate very far into the subgrade. After the stabilizing stone was applied, the stabilized soil was compacted to a high density.

A 38-mm (1.5 in.) maximum size crushed granite was used as base course. The crushed stone was placed in three 150-mm (6 in.) lifts and compacted to the maximum dry density in accordance with ASTM Test Methods for Moisture-Density Relations of Soils and Soil-Aggregate Mixtures Using 10-lb (4.54-kg) Rammer and 18-in. (457-mm) Drop (D 1557-78, Method C), as determined in the field by nuclear density tests. The total thickness of the aggregate base, including stabilizer stone, was about 533 mm (21 in.). The Los Angeles abrasion loss of the base material is 42%, and it has a magnesium sulfate soundness loss of 0.9%.

The bituminous surfacing consisted of a triple surface treatment using CRS-2H bitumen. Initially, an application of bitumen was placed on the surface of the stone base at a rate of about 0.4 gal/m^2 (0.3 gal/yd^2). This was immediately followed by an application of No. 5 stone. The application of bitumen was followed by application of No. 7 stone. Bitumen was then applied at a rate of 0.25 gal/m^2 (0.2 gal/yd^2) followed by No. 89 stone. The third application was identical to the second one.

Pavement Condition

The condition of the experimental pavement was evaluated in December 1985. The 200-m-long (650 ft.) pavement consists of a straight line segment from Station 0+00 to 1+30 (m), with an extremely sharp curve occurring from Station 1+30 to 2+00, where the pavement dead-ends at a highway.

Pavement evaluations were performed by a panel of five raters using the Georgia Department of Transportation (DOT) [*1*] and the Asphalt Institute [*2*] methods. Both approaches indicated the pavement to be in good condition, with a Present Serviceability Rating of 3.0 to 3.5. The sharp curve was excluded from the evaluation since it is not a commonly used geometric feature. From Station 0+00 to 1+30 (m), 6.5% of Class 1 and 2 longitudinal and alligator cracking had developed in the exit lane (excluding a utility trench dug across the pavement), with no cracking being visible in the entrance lane. In the straight portion of the pavement, only a very slight amount of cracking was visible in 1983 (less than 0.5%). Thus, cracking developed relatively rapidly from 1983 to 1985, indicating the need for timely maintenance for this type of pavement.

In the sharp curve near the exit to the quarry (Station 1+30 to 2+00 (m)), trucks accelerate and decelerate at the quarry entrance. Trucks also regularly go off the edge of the pavement, particularly on the inside of the sharp curve. As a result, a substantial number of 250-to

500-mm-diameter (15 to 20-in.) potholes and extensive edge ravelling had developed in this area. Some longitudinal cracking was also evident.

Traffic Loadings

The total number of fully loaded trucks at a given time having passed over the exit lane of the pavement was estimated by using observed truck distributions along with the known total quantity of stone leaving the quarry by truck each year since pavement construction. The empty truck weights were subtracted from the total allowable legal gross weight of the trucks to obtain the weight of material each truck could carry. The number of trucks was then estimated using the net weight of the material hauled, and the yearly amount of stone taken by truck from the quarry to obtain an estimated yearly total truck count.

The truck axle distributions in percent were obtained from two traffic counts taken in November 1985 to be as follows: 5%, two axles; 60%, three axles; 30%, four axles; and 5%, five axles. About 90% of the four-axle trucks had only two of the three rear axles in the lowered position while on the quarry road. On a typical day, an average of over 50 loaded trucks per hour left the quarry.

The number of equivalent 80-kN (18 kip) single-axle loads (ESAL's) was estimated using equivalency factors given by Christison et al. [*3*], Deacon [*4*], American Association of State Highway and Transportation Officials (AASHTO) [*5*], and California Department of Transportation (CALTRANS) [*6*]. The estimated ESAL's (Table 1) vary by almost an order of magnitude depending upon the equivalency factors used and whether or not the front axle loading is considered. For the experimental pavement, the total number of ESAL's through November 1985 is estimated to be approximately 3.0×10^6, including both the front and rear axles. The AASHTO equivalency factors were used for the rear axles and factors suggested by Christison et al. for the front axle loads. Neglecting front axle loadings, the total number of ESAL's is 1.4×10^6, or about one-half the actual number of equivalent axle loadings applied to the pavement by all axles present.

Rut Depths

Rut depths were measured in late February 1985 using a scale and a 1.2-m (4 ft) straight edge placed across the wheel paths. Maximum rut depths were recorded at 15-m (50 ft)

TABLE 1—*Equivalent single-axle loads through November 1985.*

No. of Axles	1985 AASHTO	CALTRANS [*2*]	Christison et al. [*3*]	Deacon [*4*]
2	18153	18153	18153	18153
3	578329	209781	953552	750922
4	749474	96635	823598	944392
55-kN front axle	...	...	19242	12689
80-kN front axle	...	...	1650929	1265712
TOTALS:				
Note 1	1375514	357432	1840116	1750251
Note 2	2653915	1635833	3118517	3028652
Note 3	3045685	2027603	3510287	3420422

NOTE 1: Totals excluding front axles.
NOTE 2: Totals including front axles determined from factors by Deacon.
NOTE 3: Totals including front axles determined from factors by Christison et al.

intervals for the entrance and exit lanes. Rut depths in the entrance lane varied from 5.1 to 8.9 mm (0.20 to 0.35 in.), with a mean of 6.9 mm (0.27 in.) and a standard deviation of 1.5 mm (0.06 in.). Rut depths in the exit lane varied from 2.0 to 15.0 mm (0.08 to 0.59 in.), with a mean of 7.4 mm (0.29 in.) and a standard deviation of 2.8 mm (0.11 in.). Hence, the more heavily trafficked exit lane had slightly higher rut depths than the entrance lane.

General Pavement Section Characteristics

Subgrade

The subgrade just below the stabilizer stone varies from a stiff, red slightly micaceous silty clay and clayey silt near the surface to a stiff to firm micaceous, slightly sandy silt at 6 m (20 ft). The standard penetration resistance (SPT-*N* value) varies from 9 to 15 and decreases slightly with depth. The groundwater table was at a depth of about 6 m (20 ft) at the time of the investigation. Good surface drainage was provided by ditches on each side of the roadway and a pavement surface gently sloping in the longitudinal direction.

The average dry density and void ratio of the subgrade was 1710 kg/m^3 (109.5 lb/ft^3) and 0.55, respectively. The moisture content of the subgrade varied from 19% to 27%. The moisture content was determined in the spring (wet season), but conditions were drier than usual at the time of the study.

An average CBR value of 10% was obtained from two unsoaked field CBR tests performed on the subgrade. Surcharge weights were applied to exert a pressure approximately equal to that of the overlying pavement.

Laboratory resilient moduli tests were performed on undisturbed, 71-mm-diameter (2.8 in.) thin-wall tube samples following previously given guidelines [7]. Resilient modulus test

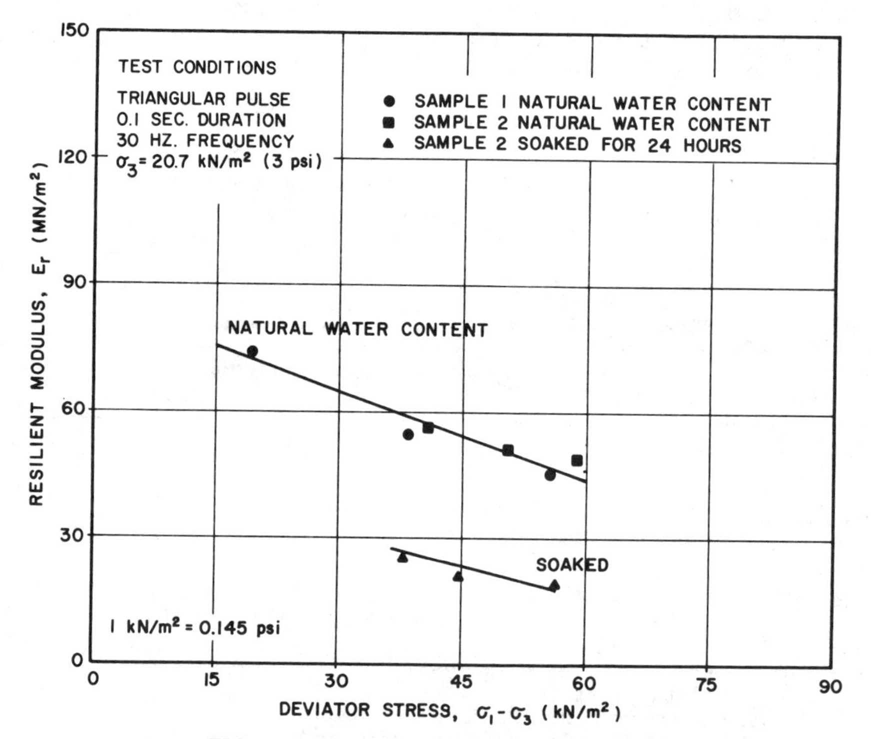

FIG. 1—*Laboratory resilient modulus of subgrade.*

conditions and results are shown in Fig. 1. After specimen conditioning, a constant confining pressure of 21 kN/m^2 (3 psi) was used in testing. Resilient moduli for the *in situ* condition ranged from 47 MN/m^2 (6800 psi) for a deviator stress of 56 kN/m^2 (8.1 psi) to 73 MN/m^2 (10700 psi) for a deviator stress of 19 kN/m^2 (2.8 psi). Soaking decreased the resilient moduli by almost one third, indicating the subgrade is quite moisture-sensitive.

Crushed Stone Base

The crushed-stone base is well graded, containing a maximum aggregate size of 38 mm (1.5 in.) with about 2.0% passing the No. 200 sieve as determined by dry sieving (Fig. 2). The subgrade stabilizer stone, as sampled, is coarser, having a similar-shaped grain size curve. Compaction tests were performed on the base coarse and the stabilizer stone sampled in 1985 in accordance with ASTM D 1557-78, Method C. The base stone had a maximum dry density of 2200 kg/m^3 (140 lb/ft^3) at a moisture content of about 6.0%. The subgrade stabilizer stone did not reach a peak maximum dry density, but had a maximum limiting value of about 2170 kg/m^3 (138 lb/ft^3). Nuclear density gage and sand cone tests indicated the existing dry base density in 1985 to 1986 to be about 2050 kg/m^3 (131.5 lb/ft^3). The average *in situ* moisture content of the base was 2.4%.

Bituminous Surface Treatment

In accordance with the ASTM Test Method for Quantitative Extraction of Bitumen from Bituminous Paving Mixes (D 2172-79) and ASTM Test Method for Recovery of Asphalt from Solution by Abson Method (D 1856-79), the following material properties were determined on specimens of the surface treatment obtained in 1985: (1) penetration of 105 at 25°C (77°F) (100 gm–5 s); (2) absolute viscosity of 1523 poises at 60°C (140°F); and (3) bitumen content of 4.1%. The surface treatment aggregate had the following gradation in percent passing: 25-mm (1 in.) sieve, 100; No. 4 sieve, 42; No. 200 sieve, 7.8.

***In Situ* Structural Strength**

Static Plate Load Tests

Field plate load tests were performed in the exit lane on the pavement surface, on the top of the base course, and on top of the subgrade (Fig. 3). A 305-mm-round (12 in.), 25-mm-thick (1 in.) smooth steel plate was loaded using a 20-ton hydraulic jack. Reaction for the jack was provided by an off-highway Caterpillar dump truck. Deflection was measured using three dial gages clamped to a 3.4-m (11 ft) aluminum beam. The beam was shaded from the sun to minimize temperature-induced movements. The plate load tests followed the procedure used by the Florida Department of Transportation (DOT), which is similar to the modified AASHTO T-22-78 Test Method.

Both mean confining stress, σ_m under full loading, and the corresponding maximum representative shearing strain γ have a significant effect on the elastic modulus of the base and subgrade [9]. In estimating the mean confining stress, both residual compaction stresses and the externally applied loading must be considered. In this study, the residual lateral stress in the granular base was taken as the full passive pressure (K_p = 7.5) corresponding to an angle of internal friction of 50 deg. The effect of σ_m and γ must be accounted for in extending the elastic moduli obtained from plate tests or other procedures to different loading conditions. For this study nondimensional relationships developed from the equations of Hardin and Drnevich [9] were used.

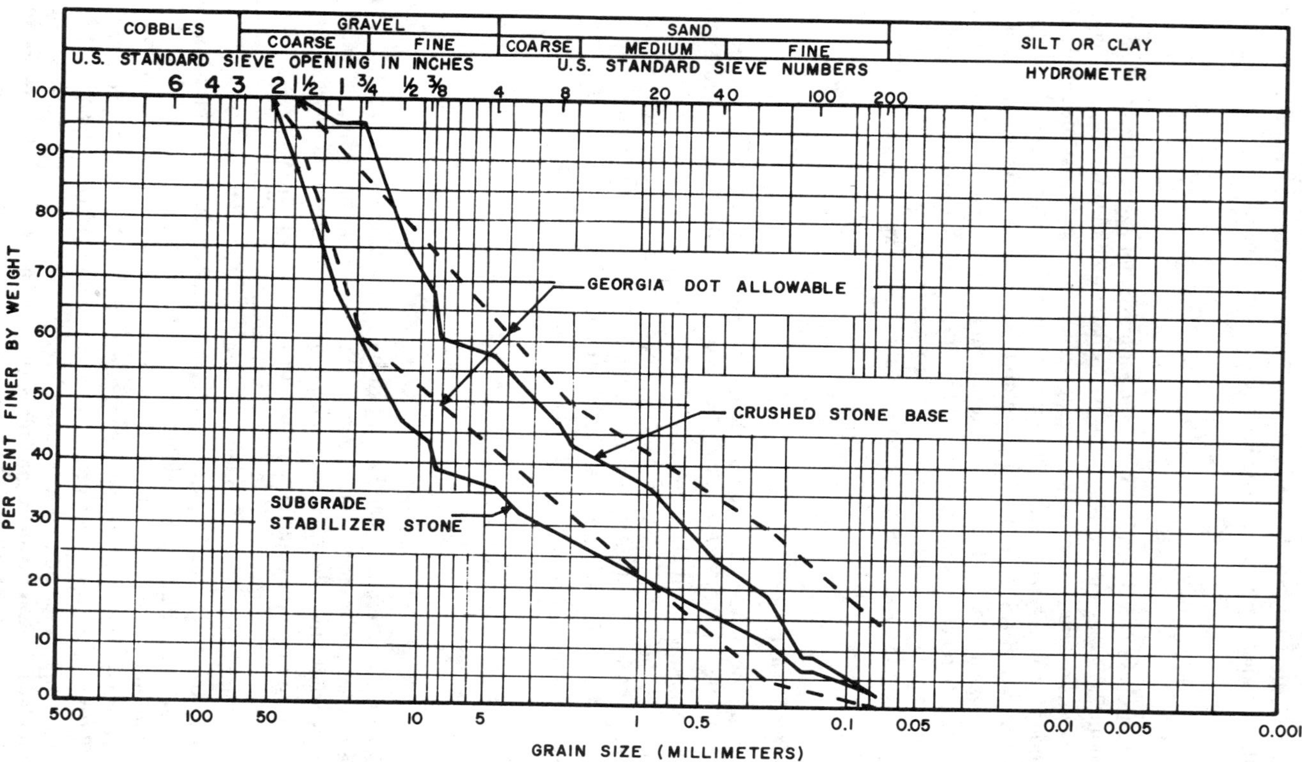

FIG. 2—*Crushed-stone base and subgrade stabilizer stone gradations.*

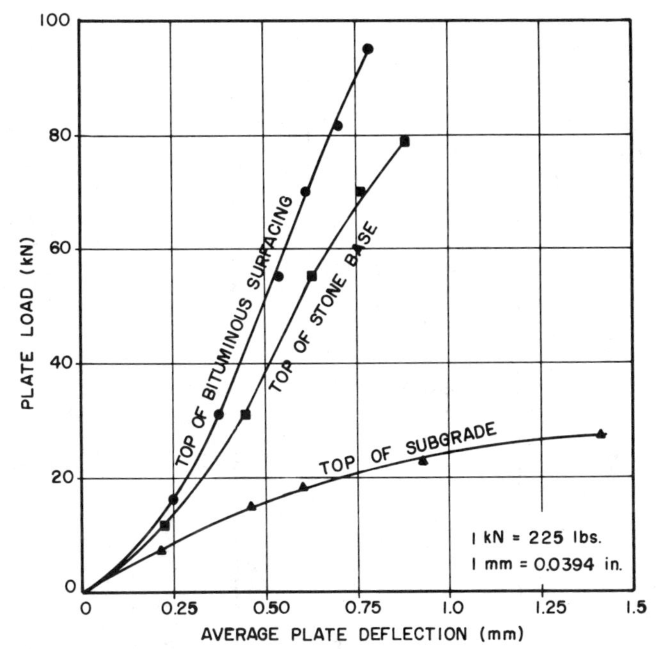

FIG. 3—*Static plate load test results.*

The elastic modulus of the subgrade and base were calculated from the plate test results using elastic half space and two-layer elastic Burmister theory, respectively. In interpreting these results, as just discussed, the correct values of σ_m and γ for each layer were considered using iteration to obtain consistent results. Calculated moduli values for the σ_m and γ conditions of the two-layer plate test are shown in Table 2. Extrapolated elastic moduli from the plate load tests are also shown for the loading conditions of the Benkelman beam and Falling Weight Deflectometer (FWD) tests. All extrapolations were iterative involving (1) estimating by experience the modulus of each layer for the desired loading condition, (2) calculating using layered theory σ_m and γ, and then the corresponding modulus of each layer, and (3) repeating Steps (1) and (2) as required until the moduli used are compatible with calculated σ_m and γ values.

Benkelman Beam Deflections and Surface Strain

Benkelman beam tests were performed using the dual wheels of an 80-kN (18 kip) single-axle dump truck in early December 1985 during a period of relatively high rainfall. On the exit lane of the pavement, deflections were measured in both the inside and outside wheelpaths at each 15-m (50 ft) station. Along the entrance lane, readings were taken alternately in the inside and outside wheelpaths at 15-m (50 ft) stations. The average pavement temperature was 13°C (56°F).

Deflections corrected to 21°C (70°F) in the exit lane ranged from 0.46 to 1.27 mm (0.018 to 0.05 in.), with an average of 0.70 mm (0.0275 in.) for the inside wheelpath and 0.71 mm (0.0281 in.) for the outside wheelpath; one standard deviation was 0.18 and 0.20 mm (0.007 and 0.008 in.), respectively. Deflections in the entrance lane ranged from 0.41 to 0.84 mm

TABLE 2—*Base and subgrade moduli.*[a]

	Plate Test			Benkelman Beam				FWD—40-kN Load				
				Extrapolated Plate E				Extrapolated Plate E				
Layer	σ_m, kN/m^2	γ[b]	E, MN/m^2	σ_m, kN/m^2	γ[b]	E, MN/m^2	Back-calculated E, MN/m^2	σ_m, kN/m^2	γ[b]	E, MN/m^2	WES E, MN/m^2	University of Illinois E, MN/m^2
Base	87.5	0.027	210	98.5	0.25	110	63	101.2	0.06	214	406	...
Subgrade	44.0	0.06	71[c]	24.8	0.06	36	45	24.8	0.037	55	220	117

[a] 1 psi = 6.9 kN/m^2; 1 ksi = 6.9 MN/m^2; 1 kip = 4.5 kN.
[b] Shear strain γ in percent.
[c] Calculated for plate test on top of base by correcting half-space modulus considering subgrade σ_m and γ.

(0.016 to 0.033 in.), with an average of 0.64 mm (0.025 in.) for both the inside and outside wheelpaths. Deflections in the more heavily loaded exit lane, therefore, were generally higher than those in the entrance lane.

Both longitudinal and transverse surface strains were measured in the exit lane using the same dual-wheel, 80-kN (18 kip) axle loading as for the Benkelman beam tests. These tests were performed close to where plate load tests and FWD profile tests were performed. Wire resistance strain gages were used having an active gage length of 51 mm (2 in.). To install the gages, small slots were cut in the bituminous surface about 13 mm (0.5 in.) wide and 3 mm (0.125 in.) deep using a router having a carbide bit and a special steel guide frame. A fully temperature-compensated, four-leg Wheatstone bridge configuration was used for the strain measurements.

Measured longitudinal compressive strains at a pavement temperature of 13°C (56°F) varied from about 80 to 114 $\times 10^{-6}$ for truck speeds of 5 to 8 km/h (3 to 5 mph); transverse strains varied from 122 $\times 10^{-6}$ to 158 $\times 10^{-6}$. Mean values of the recorded strains have little meaning since the maximum strain occurs only when the dual tires are positioned perfectly over the gages. Therefore, for analysis purposes the average of only the larger observed strains was used giving 112 $\times 10^{-6}$ and 150 $\times 10^{-6}$ for longitudinal and transverse strains, respectively.

Back-Analysis of Moduli From Benkelman Beam Tests

The Chevron 5-Layer computer program with capability for nonlinear material properties was used to backcalculate the effective moduli of the base and subgrade from the Benkelman beam and strain measurement results. In the calculations, the base was assumed to consist of three 150-mm (6 in.) layers and have elastic moduli E that varies as

$$E = K\sigma_\theta^n$$

where σ_θ is the sum of the principal stresses and K and n are material properties.

The constant n was taken to be a fixed value of 0.8. The constant K was adjusted to give a best fit between the calculated and observed values of maximum surface strains and average Benkelman beam deflection between the dual wheels.

The bituminous surfacing was estimated to have a modulus of 6890 MN/m^2 (1 $\times 10^6$ psi) [*10*]. Based on the results of the plate load tests on the subgrade and the repeated load triaxial tests, the modulus of the subgrade was taken to be 4.5 MN/m^2 (6500 psi). The nonlinear, iterative Chevron 5-layer program gave an average base course elastic modulus of 6.3 MN/m^2 (9200 psi), with the modulus decreasing slightly in the base with increasing depth. Longitudinal and transverse compressive strains were calculated of 106 $\times 10^{-6}$ and 133 $\times 10^{-6}$, respectively, which compare reasonably well with the measured values.

Falling Weight Deflectometer Tests

Nondestructive tests were conducted by the Waterways Experiment Station (WES) on the road using the Dynatest Model 8000 Falling Weight Deflectometer (FWD). The Dynatest 8000 FWD is an impact load device that applies a single-pulse, transient load of 25 to 30 ms duration. The applied force and pavement deflections are measured with load cells and velocity transducers. The drop height can be varied from 0 to 400 mm (15.7 in.) to produce a force from about 6 to 110 kN (1.5 to 25 kips). The load is transmitted to the pavement through a 300-mm-diameter (11.8-in.) plate. The system is controlled with a Hewlett-Packard HP-85 computer, which also records the output data.

FWD data collected were deflection basin measurements for force levels of about 40, 70, and 110 kN (9, 15, and 25 kips). Displacements were measured at the center of the load plate and at distances of 0.30, 0.61, 0.91, 1.22, 1.52, and 1.83 m (12, 24, 36, 48, 60, and 72 in.) from the center of the plate. FWD tests were performed at 15-m (50 ft.) intervals on four lanes of the experimental pavement: Lanes 1 and 2 were along the outside and inside wheel path of the exit lane from the quarry; Lanes 3 and 4 were along the inside and outside wheel path of the entrance lane. An Impulse Stiffness Modulus (ISM) profile of the experimental road for each lane is shown in Fig. 4. Impulse stiffness modulus is defined as the FWD load divided by the deflection at the center of the load. ISM data obtained on Lane 1, the outside wheel path of the heavily loaded exit lane, are much lower than for the other three lanes. The ISM stiffness at different FWD force levels indicate if a pavement is stress-hardening or stress-softening. The average stiffness for each lane of the thick crushed-stone base pavement increases with the FWD force level as shown in Fig. 5, indicating a stress-hardening type pavement structure.

Backcalculated Moduli

The BISDEF computer program developed at WES was used to backcalculate moduli values from the FWD deflection basin data [*11*]. BISDEF uses the BISAR layered elastic program to calculate deflections to match the seven measured deflections of a FWD basin. For these calculations, the pavement was considered to consist of a 559-mm (22 in.) base over a subgrade. A rigid layer was placed at a depth of 6.1 m (20 ft) below the pavement surface. Representative basins were selected for each lane tested. Backcalculated moduli for Lanes 2 through 4 and a 40-kN (9-kip) load were about 406 MN/m^2 (59 ksi) for the base and 220 kN/m^2 (32 ksi) for the subgrade. As the load level increased from 40 to 110 kN (9 to 25 kips), the backcalculated modulus increased by about 56%, with the subgrade modulus only increasing slightly. For Lane 1, the modulus of the base was about the same as the reported value for the subgrade.

The moduli calculated using BISDEF from FWD data by WES are compared in Fig. 6 to those of airfield pavements (Table 3) recently tested. The backcalculated base course

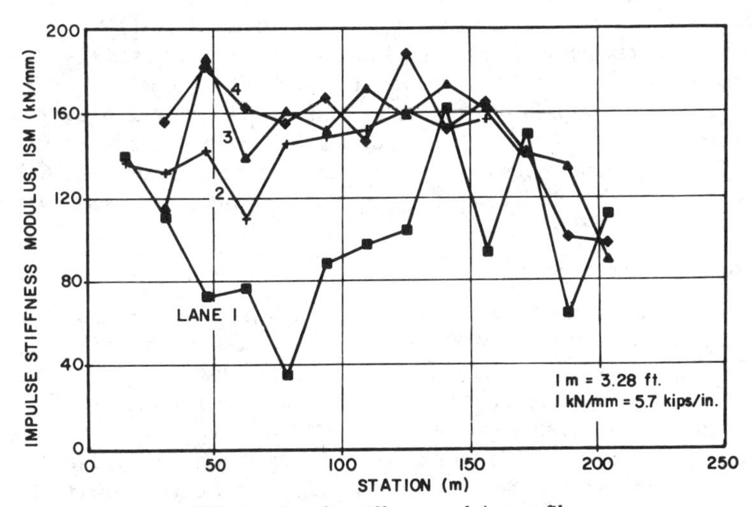

FIG. 4—*Impulse stiffness modulus profiles.*

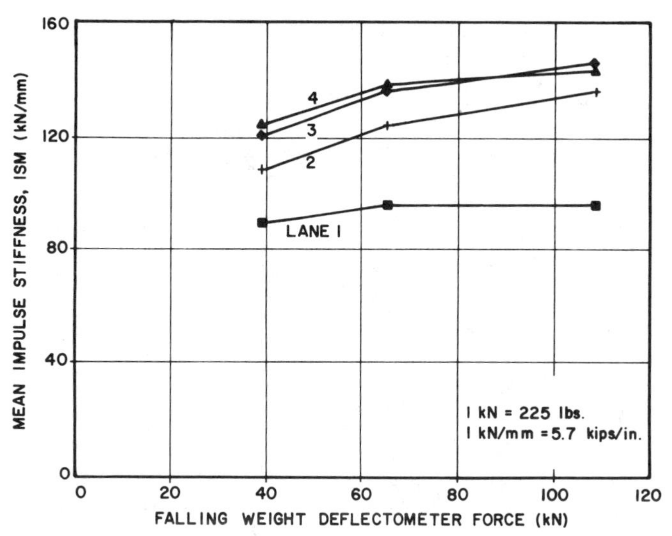

FIG. 5—*Impulse stiffness as a function of force.*

modulus for the Stockbridge test pavement is considerably higher than those from other test sites having comparable subgrade moduli.

Also, R. L. Lytton analyzed the Stockbridge deflection basin data using six different methods recommended by Lytton et al. [*12*]. The pavement was assumed to consist of 19 mm (0.75 in.) of bituminous surfacing, 533 mm (21.0 in.) of aggregate base, and an infinite subgrade. The elastic moduli of the base and subgrade obtained by Lytton are given in Table 4.

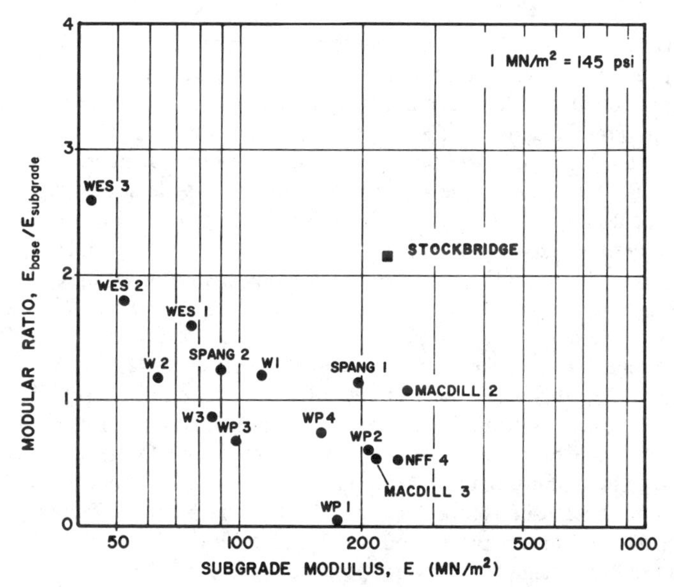

FIG. 6—*FWD relationship between subgrade modulus and modular ratio.*

TABLE 3—*FWD pavement properties.*[a]

	Surface		Base			Subgrade		
Site	Type	Thickness, mm	Type	Thickness, mm	CBR	Type	CBR	Age, Years
WES1	AC	43	GW	208	100	CH	7	0
WES2	AC	36	GW	229	100	CH	6	0
WES3	DBST	13	GW	239	100	CH	6	0
WP1	AC	76	GW	152	12	...	...	22
WP2	AC	76	GP-GC	1194	33	...	...	24
WP3	AC	50	GP	305	33	SC	7	9
WP4	AC	50	GW-GM	305	72	SC	8	21
W1	DBST	25	GC	737	33	...	...	22
W2	AC	64	GC	305	102	CH	4	30
W3	AC	64	GC	406	37	CL	4	24
NFF4	AC	53	GW	160	100	SP-SM	20	0
SPANG1	AC	51	GP-GM	156	80	CL	11	0
SPANG2	AC	89	GP-GM	420	100	CL	...	0
MACDILL2	AC	254	SM	381	80	SP	30	41
MACDILL3	AC	140	SM	381	80	SP	30	41

[a] 1 in. = 25 mm.

Deflection basins were also analyzed by Thompson using the nonlinear ILLI-PAVE generated algorithms [*13*] for Lanes 1 through 4. This back-analysis technique indicated a subgrade elastic modulus of about 117 MN/m^2 (17 ksi) for each lane.

Discussion

Base Course Layer Coefficients

The effective base course layer coefficients for use in the AASHTO Interim Guide [*14*] method of pavement design can be evaluated using the known load applications and the structural characteristics of the pavement. Front-axle loads are commonly not considered in evaluating pavements and hence were not included in this analysis.

A soil support value of 4.75 is appropriate for the measured subgrade CBR of 10 [*14*]. In wet springs, the CBR and hence soil support value may be less. A regional factor of 1.8 would usually be used for Atlanta by the Georgia DOT. The following thicknesses and layer coefficients, a_i were also used in the analysis: surface = 19 mm (0.75 in.), a_1 = 0.44; upper portion of base = 286 mm (11.25 in.), a_2 = 0.18; lower portion of base = 248 mm (9.75 in.), a_3 = unknown. Use of a_1 = 0.44 and a_2 = 0.18 for the upper 305 mm (12.0 in.) of pavement is in agreement with Georgia DOT practice. Reducing the value of a_i below 305 mm (12.0 in.) considers that the stone (or any other base material) is not as effective with increasing depth [*15*]; this is also true of any other type base materials [*15*].

Using the above structural properties, a known traffic loading of 1.4 × 10^6 ESAL's, and a present serviceability rating of 3.0, the unknown lower crushed stone base layer coefficient was calculated to be 0.16 using the Interim Guide method [*13*]. The layer coefficient of the upper portion of the crushed-stone base was 0.18 in the analysis. Thus, the Stockbridge crushed-stone base pavement performed considerably better than indicated by layer coefficients often used for design [*14*]. Several possible reasons for the good performance of the pavement include (1) use of a 38-mm (1.5 in.) maximum size, subangular crushed granite

TABLE 4—*Base and subgrade elastic moduli backcalculated from FWD deflection basin results.*

	Elastic Modulus, MN/m²[a]			
	Lane 1		Lane 2	
Method	Base	Subgrade	Base	Subgrade
ELSDEF	313	249	333	332
CHEVDEF[b]	254	281	341	324
NEWSER[c]	...	...	385	208
MODULUS	349	629	506	689
MODCOMP2[d]	164	340	164	340
BISDEF[b,e]	256	262	336	314
Average	269	351	345	365

[a] 1 ksi = 6.9 MN/m².

[b] Surface layer calculations for CHEVDEF and BISDEF had to be "fixed" within the min-max range to obtain values for the base course and subgrade.

[c] NEWSER2 would not calculate Lane 1 data set; surface layer thickness is not within the range of application; results are extrapolated; subgrade modulus is probably most accurate.

[d] MODCOMP2 was developed by L. H. Irwin (copyright by Cornell University).

[e] BISDEF was developed by the Waterways Experiment Station. BISDEF contains the proprietary Bisar Layered Elastic Program.

with a low percent fines (the dry gradation tests performed on *in situ* material indicated 2% fines), (2) presence of a good subgrade, and (3) good general drainage and a low condition of moisture in the base. Also, the thin-surface treatment gives a very flexible pavement with good fatigue resistance.

Equivalent Axle Loads

Equivalent axle loads vary significantly depending upon the method used for evaluation. The approach of Christison et al. [*3*] appears to be most appropriate (and conservative) for cases in which the primary failure mechanism is fatigue since the method is based on measured pavement surface strains. Finally, front axle loads should be considered in design methods since they can have an important detrimental effect on the pavement life.

Elastic Modulus

The elastic modulus E for a given material is strongly dependent upon the mean confining pressure σ_m and the shearing strain γ, and is effected to a lesser degree by factors such as plasticity and over-consolidation stress [*9,16*]. In the past, only the sum of the principal stresses (which is analogous to using σ_m), has been commonly used to define the elastic modulus. This study indicates that the shearing strain γ or deviator stress [*17*] should also be considered for an accurate estimate of the elastic modulus.

In general, under a plate loading, because of differences in σ_m and γ as indicated in Table 2, the elastic modulus observed appears to be higher than for a typical dual-wheel loading. Therefore, care should be exercised in extrapolating either conventional plate test results or FWD results to dual-wheel-type pavement loadings.

As shown in Table 2, the estimated elastic modulus of the crushed-stone base and subgrade at Stockbridge backfigured from FWD deflection basins is 406 MN/m² (59 ksi) and 220 MN/

m^2 (32 ksi), respectively, by the BISDEF method [*11*]. This value of modulus is reasonably close to the average base value predicted by Lytton (Table 4). Extrapolation of the static plate load test results indicates an elastic modulus for the base of about 214 MN/m^2 (31 ksi) and 55 MN/m^2 (8 ksi) for the subgrade. The static plate test results (Table 2) contain effects of creep which result in lower estimated values of moduli, particularly for the subgrade. This at least partially explains why the FWD subgrade moduli estimated from the static plate load tests were so low, since the effects of creep become greater as plasticity increases [*9*]. In contrast, the static plate tests gave relatively good results for the Benkelman beam tests, where some creep effects were present. Hence, the moduli from the Benkelman beam tests are lower than would be expected for a moving wheel load.

Under a dual-wheel loading, a consideration of σ_m and γ indicates that the elastic modulus of the crushed-stone base, extrapolated from the FWD results, might be on the order of 210 to 280 MN/m^2 (30 to 40 ksi). *In situ* measurements made in South Africa [*18*] show the elastic modulus of a thin, relatively dry, very dense granular base on top of a rigid layer to be about 400 MN/m^2 (58 ksi) for an 80-kN (18-kip), dual-wheel loading and about one million ESAL's. The crushed-stone base was dense graded, with a 38-mm-diameter (1.5 in.) maximum size crushed stone. Because of the special base characteristics (thin thickness, high density, and underlying rigid layer) it is possible this modulus is approaching the maximum likely value for the size and gradation stone used in a reasonably dry condition under a moving 80-kN (18-kip), dual-wheel loading. Van Zyl and Maree [*18*] also observed almost a 50% reduction in base elastic modulus in going from a relatively dry to an almost saturated condition. Their results indicate the important effect which moisture can have on base performance.

The ILLI-PAVE method gave an elastic modulus of the subgrade of 117 MN/m^2 (17 ksi), which is close to one-half the value predicted by BISDEF. One possible explanation for this low value is the fact that the average vertical stress in the subgrade, because of the thick granular base, was much less than the 441 kN/m^2 (6 psi) assumed in the analysis. Because of the use of nonlinear subgrade properties, this assumption could have resulted in an under-prediction of E.

At the present time, a relatively wide variation in moduli values may be obtained using different methods of analysis, as indicated in Table 4. A need exists, therefore, for further research in backcalculating elastic moduli from FWD deflection basin data. The incorporation of nonlinear material properties into the analysis would be quite desirable. Verification of the methods need to be carried out using additional measured pavement response variables such as deflections of each layer, strains in the bituminous layer, and vertical stresses.

The elastic modular ratio between a stone base and subgrade (Table 2 and Fig. 6) appears from the FWD tests as interpreted by WES to vary between about 1.0 and 2.5 for bases having CBR values greater than about 100. For the Stockbridge test road, this ratio was about 2.1 for corresponding loads. The modular ratio observed at Stockbridge is within the range usually used for design as summarized by Smith and Witczak [*19*].

Pavement Strength Variation

The FWD results (Fig. 4) show that the strength of the pavement varies significantly with the lane. This finding indicates that the pavement strength (1) decreases with load level to which the pavement has been subjected and (2) decreases as the distance of load from the edge of the pavement decreases. The Benkelman beam tests also showed this same general trend, but to a much lesser degree. The difference in severity of these effects with the test method contributes further evidence that the response of the pavement varies with the test method, and caution should be exercised in extrapolating test results.

Conclusions

The crushed-stone base at the Stockbridge test road demonstrated excellent performance under large numbers of heavy loads. Backcalculated AASHTO layer coefficients of the upper 286 mm (11.25 in.) and lower 248 mm (9.75 in.) of the base are 0.18 and 0.16, respectively. The elastic modulus of the base and subgrade vary with the loading condition and method of data reduction. The mean confining stress and shearing strain must both be considered in translating moduli from one loading condition to the other. For static loading, creep effects are also important. Finally, more research is needed in analyzing nondestructive test data in evaluating pavement strength.

Acknowledgments

The authors wish to thank R. L. Lytton for the comprehensive study of the moduli calculated by different methods and M. R. Thompson for calculating the FWD elastic moduli using ILLI-PAVE algorithms. The authors gratefully acknowledge the support of the U.S. Army Engineers Waterways Experiment Station and Vulcan Materials Company during the testing and reporting phase of this study.

References

[*1*] "PACES, Pavement Condition Evaluation System," Georgia Department of Transportation, June 1983.
[*2*] "Asphalt Overlays for Highway and Street Rehabilitation," Manual Series No. 17, Asphalt Institute, June 1983.
[*3*] Christison, J. T., Anderson, K. O., and Shields, B. P. *Proceedings,* Association of Asphalt Paving Technologists, Vol. 47, 1978, pp. 398–433.
[*4*] Deacon, J. A., *Proceedings,* Association of Asphalt Paving Technologists, Vol. 38, 1969, pp. 465–491.
[*5*] "Proposed AASHTO Guide for Design of Pavement Structures," Project 20-7/20, National Cooperative Highway Research Program, Washington, DC, March 1985.
[*6*] "Highway Design Manual," CALTRANS, California Department of Transportation, Sacramento, California, 1981, pp. 7-651.1–7-651.6.
[*7*] "Test Procedures for Characterizing Dynamic Stress-Strain Properties of Pavement Materials," Transportation Research Board, Special Report 162, 1975.
[*8*] Barksdale, R. D., *Proceedings,* American Society of Civil Engineers, Vol. 103, TE1, Jan. 1977, pp. 55–73.
[*9*] Hardin, B. O. and Drnevich, V. P., *Proceedings,* American Society of Civil Engineers, Vol. 98, SM7, July 1972, pp. 667–692.
[*10*] Shook, J. F., Finn, F. N., Witczak, M. W., and Monismith, C. L. in *Proceedings,* 5th International Conference on the Structural Design of Asphalt Pavements, Vol. 1, The Netherlands, Aug. 1982, pp. 17–44.
[*11*] Bush, A. J. and Alexander, D. R., Research Record 1022, paper presented at Transportation Research Board, Washington, DC, Jan. 1985, pp. 16–28.
[*12*] Lytton, R. L., Roberts, F. L., and Stoffels, S., National Cooperative Highway Research Program Final Report 10–27, April 1986.
[*13*] Thompson, M. R. and Elliot, R. P., Research Record 1043, Transportation Research Board, 1985, pp. 50–57.
[*14*] "AASHTO Interim Guide for Design of Pavement Structures," American Association of State Highway and Transportation Officials, Washington, DC, 1974.
[*15*] Barksdale, R. D., "Performance of Base Courses," American Society of Civil Engineers Specialty Conference on Solutions for Pavement Rehabilitation Problems, Atlanta, May 1986, pp. 97–112.
[*16*] Brown, S. F. and Pappin, J. W., Research Record 1022, Transportation Research Board, 1985, pp. 45–51.
[*17*] Uzan, J., Research Record 1022, Transportation Research Board, 1985, pp. 52–59.
[*18*] Van Zyl, N. J. W. and Maree, J. H., *Civil Engineer in South Africa,* July 1983, pp. 365–376.
[*19*] Smith, B. E. and Witczak, M. W., *Proceedings,* American Society of Civil Engineers, Vol. 107, TE6, Nov. 1981.

Elton R. Brown,[1] John L. McRae,[2] and Alfred B. Crawley[3]

Effect of Aggregates on Performance of Bituminous Concrete

REFERENCE: Brown, E. R., McRae, J. L., and Crawley, A. B., **"Effect of Aggregates on Performance of Bituminous Concrete,"** *Implication of Aggregates in the Design, Construction, and Performance of Flexible Pavements, ASTM STP 1016,* H. G. Schreuders and C. R. Marek, Eds., American Society for Testing and Materials, Philadelphia, 1989, pp. 34–63.

ABSTRACT: Because aggregate comprises approximately 95% of asphalt mixtures, it has a major effect on the performance of mixtures. The quality of filler and amount of filler greatly affect asphalt concrete performance. To insure satisfactory performance, procedures must be available that can be used to detect poor-quality mixtures and there must be techniques available for improving mixture properties.

The objective of this paper is to present data from various studies that show the effect of aggregate grading on performance of asphalt mixtures. Test methods to evaluate these mixture properties are discussed.

A well-graded, crushed aggregate should be used to provide the highest-quality asphalt concrete. Uncrushed aggregates such as natural sands and uncrushed gravels produce mixtures with lower stability and decreased durability, whereas a well-graded aggregate fits together more tightly during compaction, resulting in a lower required asphalt content and improved stability and durability.

The maximum aggregate size is important to control stability, skid resistance, and compactibility of the mixture. A larger maximum size aggregate produces a higher stability and usually requires less asphalt content. The larger size aggregate also provides for improved skid resistance. Compaction of thin layers is difficult when the maximum aggregate size is too small. The compacted lift thickness should be at least twice the maximum aggregate size.

The amount of mineral filler in a mixture greatly affects the overall mix quality. Since the filler tends to fill the voids in the mineral aggregates, an increase in filler results in a corresponding decrease in optimum asphalt content. Laboratory test results also indicate that a mixture design performed for an aggregate with higher filler content will result in a significantly higher stability than that for lower filler content. The type of filler used also affects the mixture properties.

The quality of the aggregate must be controlled to insure a quality asphalt concrete. Some properties that must be considered are maximum aggregate size, amount of crushed particles, and amount and quality of filler.

KEY WORDS: asphalt, bituminous concrete, compaction, density, filler, gradation, Gyratory Testing Machine (GTM), index, plasticity, rutting, shear, size, skid resistance, stability, stripping tendencies, voids in mineral aggregate (VMA)

In recent years, increases in tire pressures and in volume of traffic have been a significant factor in increased costs for maintenance and repair of asphalt concrete pavements. This higher cost has emphasized the importance of selecting high-quality materials and designing and placing these bituminous mixtures using the latest technology.

[1] Research civil engineer, Waterways Experiment Station, P.O. Box 631, Vicksburg, MS 39180.

[2] Consultant, president, EDCO Inc., P.O. Box 1109. Vicksburg, MS 39180.

[3] Assistant research and development engineer, Mississippi State Highway Department, P.O. Box 1850, Jackson, MS 39215.

Of major concern is the quality and gradation of the aggregate, which makes up approximately 95% of the mix by weight. Aggregates have been studied over the years, but performance of bituminous mixtures is difficult to model using aggregate properties such as gradation, hardness, toughness, porosity, or other related properties.

This paper includes some of the experiences of the U.S. Army Corps of Engineers and the Mississippi and Louisiana State Highway Departments with asphalt concrete performance as it relates to aggregate properties. This paper discusses the effect of mineral filler, maximum aggregate size, aggregate gradation, crushed particles, and stripping tendencies on performance of asphalt concrete. Improved methods are recommended for evaluating mixtures to determine overall optimum conditions and overall performance with particular concern for the aggregate phase.

Effect of Filler Type

More than 30 years ago the Corps of Engineers used asphalt mixtures containing loess filler in riverbank stabilization. At that time, test results indicated that the loess mixtures were more impervious to water than other types of asphalt mixtures. Because of the improved ability of loess to provide a waterproof layer, and other possible uses, a limited study was initiated to look at the potential use of loess in sand asphalt mixtures for pavements.

A plan of tests was developed to compare sand asphalt mixtures containing loess filler with sand asphalt mixtures containing limestone dust filler. Sand mixtures were evaluated with no filler and with 4, 8, and 12% of each filler type. The bitumen content which provided 6% voids total mix in the limestone dust mixes was used to prepare all specimens, although the bitumen content at 6% voids total mix in the loess mix would have been slightly different.

Sample preparation was performed with a Gyratory Testing Machine (GTM) with the normal or vertical pressure set at 689 kPa (100 psi) and with a 1-deg angle and 30 revolutions. This is closely equivalent to 50-blow Marshall compactive effort. Twelve samples of each mix type were prepared for determining unit weight, voids, stability, flow, retained stability (vacuum saturation), and indirect tensile strength. All tests were conducted in acccordance with Military Standard 620A with the exception of retained stability after vacuum saturation and indirect tensile strength, which were both conducted in accordance with proposed standards under jurisdiction of ASTM Subcommittees D04.22 and D04.20, respectively.

Physical properties of the sand, limestone dust, loess, and the AC-20 asphalt cement are shown in Table 1. A list of the seven mixtures and the asphalt content used for preparing mixtures for this study is shown in Table 2.

TABLE 1—*Material properties.*

Sand Gradation				
Sieve Size	Percent Passing	ASTM Apparent Specific Gravity		Asphalt
⅜ in. (0.95 cm)	100	limestone dust	2.72	specific gravity, 1.032
No. 4	92	loess	2.67	penetration [0.1 mm (0.0039 in.)], 69
No. 8	81	sand	2.65	viscosity at 60°C (140°F), 2089 poises
No. 16	74			viscosity at 135°C (275°F), 492 cSt
No. 30	59			
No. 50	11			
No. 100	3			
No. 200	2.4			

TABLE 2—*Mix identification.*

Mix	Description
A	aggregate: 100% sand; 9.6% AC-20 asphalt
B	aggregate: 96% sand, 4% limestone dust; 8% AC-20 asphalt
C	aggregate: 92% sand, 8% limestone dust; 6.6% AC-20 asphalt
D	aggregate: 88% sand, 12% limestone dust; 5.5% AC-20 asphalt
E	aggregate: 96% sand, 4% loess; 8% AC-20 asphalt
F	aggregate: 92% sand, 8% loess; 6.6% AC-20 asphalt
G	aggregate: 88% sand, 12% loess; 5.5% AC-20 asphalt

A summation of the average test results for each of the seven mixtures is shown in Table 3. A review of the density results shows that an increase in loess filler actually decreased the mix density slightly, while an increase in limestone filler increased the density significantly (Fig. 1*a*). This indicated that the limestone filler fills the voids in the sand mix while the loess filler for some reason bulks the sand and thus prevents densification. This is likely to be related to the largely unexplored effects of body forces of the fine particles; that is, basic considerations relating to the balance between the applied stress and the aggregate particle combined body forces of repulsion and attraction, including the effects of bituminous films surrounding the particles [*1–3*].

The relationship between voids in mineral aggregate (VMA), which is the inverse function of unit weight aggregate only, and filler content is another way of showing the effect of filler on compaction (Fig. 1*b*). Note that the VMA percentage varies from 26 to 18% as the filler content varies from 0 to 12% for the mix containing limestone filler, while the VMA only varies from 26 to 23% as the filler content varies from 0 to 12% for the mix containing loess filler.

The data (Fig. 1*c*) show that the stability increases with increasing filler contents. The stability increase is greater for the limestone filler than for the loess filler. This greater increase in stability for limestone filler mixtures is expected as a result of the greater increase in density for these mixtures as long as the increased density does not create a condition of overfilled voids [*1,4–7*].

TABLE 3—*Mix properties.*

Mix Type	Density, lb/ft³[a]	VMA, %	Stability, lb[b]	Retained Stability After Static: Immersion, lb	(%)	Retained Stability After Vacuum: Saturation, lb	(%)	Flow, 0.01 in.[c]	Tensile Strength, psi[d]
A	135.7	25.7	216	220	(102)	13	(6)	12	47.0
B	137.3	23.3	273	300	(110)	88	(32)	14	54.2
C	140.7	20.3	534	541	(102)	207	(39)	12	69.4
D	142.8	18.1	833	756	(91)	313	(38)	11	80.6
E	135.4	24.4	250	240	(96)	53	(21)	14	53.4
F	135.3	24.3	360	283	(79)	88	(24)	11	59.2
G	135.3	22.5	514	420	(82)	92	(18)	10	61.8

[a] Lb/ft³ × 16.01 = kg/m³.
[b] Lb × 0.45 = kg.
[c] In. × 2.54 = cm.
[d] Psi × 6.894 = kPa.

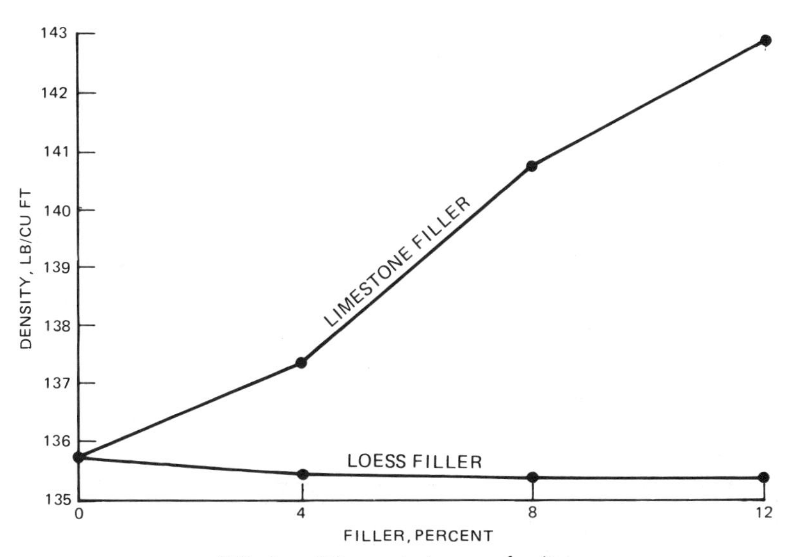

FIG. 1*a*—*Filler content versus density.*

Two types of water susceptibility tests were conducted; one was the static immersion used by the Corps of Engineers and the other was a vacuum saturation method proposed as a standard under the jurisdiction of ASTM Subcommittee D04.22. The static immersion tests on these samples showed the values of retained stability to range from 79% for Mix F to 110% for Mix B. The criterion used by the Corps of Engineers is 75% minimum retained stability, and every mix met this requirement. The mixes containing limestone filler generally performed better than the mixes containing loess.

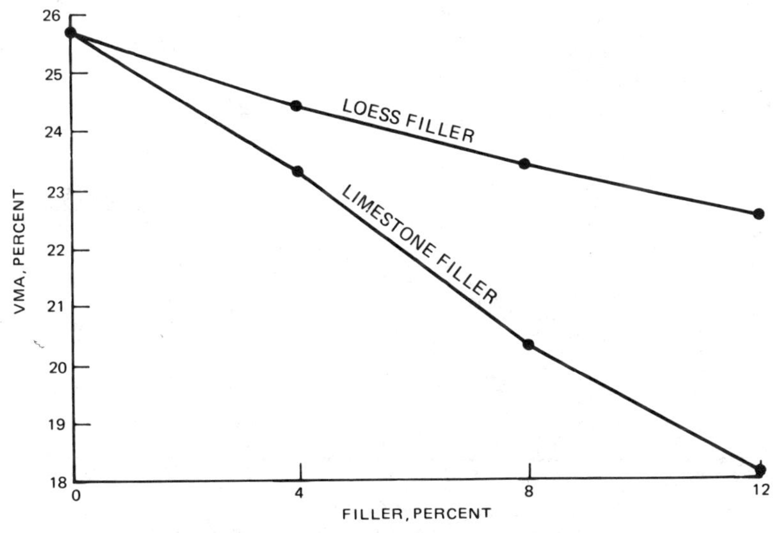

FIG. 1*b*—*Filler content versus VMA.*

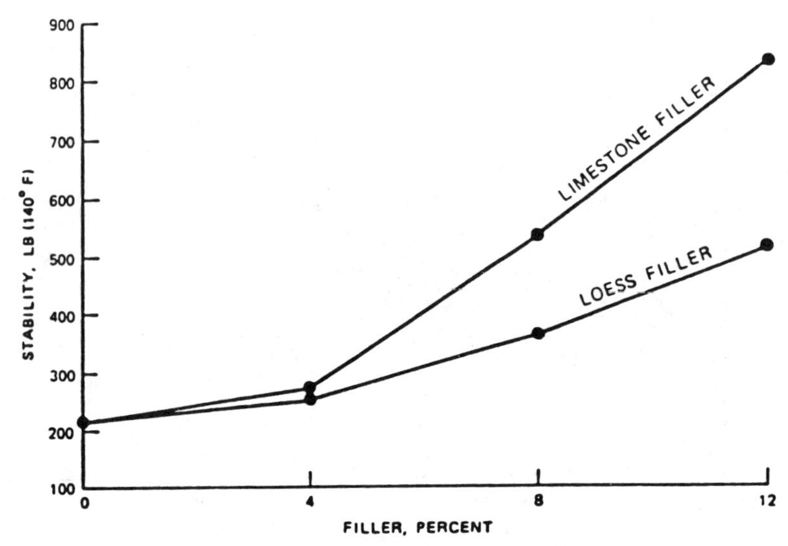

FIG. 1*c*—*Filler content versus stability.*

The vacuum saturation test is more severe than the static immersion test. There is no universally accepted criterion developed for this method, but 50% minimum retained stability is often used. This test appears to divide the mixes into three categories. Mix A, which contained 100% sand, had the lowest retained stability of all mixes (6%). The three mixes prepared with loess material had retained stabilities of 21, 24, and 18%, indicating some improvement over the sand mix. The mixes prepared containing limestone dust, on the other hand, had retained stability of 32, 39, and 38%, showing considerable improvement over the mixes containing loess.

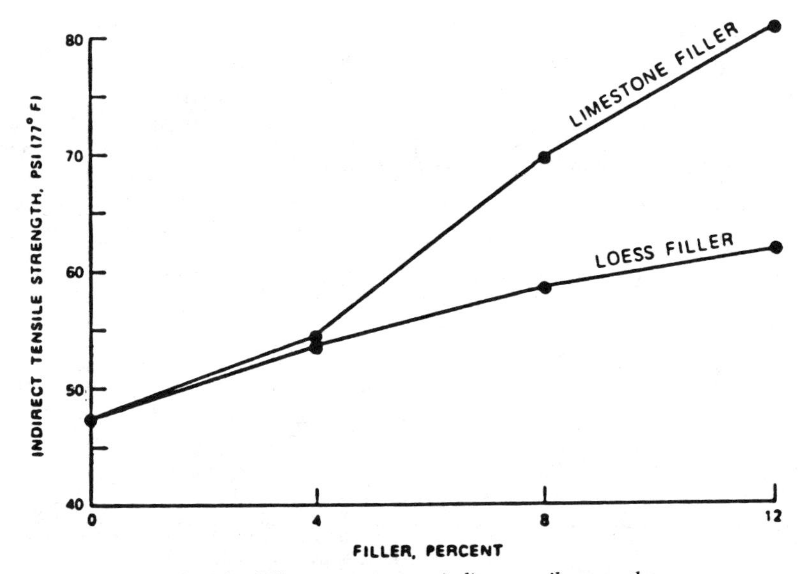

FIG. 1*d*—*Filler content versus indirect tensile strength.*

The tensile strength test results show that the use of limestone filler increases the tensile strength more than the increase provided when loess filler is used (Fig. 1*d*). The indirect tensile strength results show the same trends as those shown by the stability tests.

The stability and indirect tensile strength for all seven mixes were plotted as a function of VMA in Fig. 2*a* and 2*b*, respectively. These plots indicated that the stability and indirect tensile strength are closely related to VMA, which is inversely related to unit weight aggregate only. This supports the fact that the difference in compactability is the major reason for differences in stability and indirect tensile strength properties.

The retained stability after vacuum saturation does not appear to be related to density, but rather to properties of mineral filler as indicated earlier. Lime has been used for a number of years as an antistrip agent; therefore, it is reasonable to expect that the use of limestone filler will reduce the water susceptibility of the sand mix.

Effects of Filler Amount

A laboratory study was set up to evaluate the effect of varying amounts of −200 material on the quality of asphalt concrete. Limestone aggregate was blended to meet the Corps of Engineers gradation requirements for ¾-in. (1.9-cm) maximum size aggregate. The amount of filler (limestone dust) was then varied to produce mixes having 3, 6, 8, 10, and 12% −200 material. The mixture containing 8, 10, and 12% −200 material did not meet the Corps of Engineers criteria for gradation. Each of these aggregate gradations was mixed with amounts of asphalt varying from 4.0 to 7.0% in 0.5% increments to determine optimum asphalt content. The results of laboratory tests conducted on these samples are shown in Figs. 3 to 7.

The optimum asphalt content for the mix containing 3% −200 material was selected to be 6.6%. Table 4 compares Corps of Engineers criteria for high-pressure-tire design pavements with laboratory-determined values for a mix with 6.6% asphalt content and 3% −200 material.

It can be seen that two of the criteria (stability and voids filled with asphalt) are not met. The only way to bring the voids filled with asphalt (Fig. 5) within the specified range is to decrease the asphalt content, and the only way to bring the stability value (Fig. 6) above the minimum allowed is to increase the asphalt content. Hence, this particular aggregate gradation would not produce a mix which met Corps of Engineers criteria.

The optimum asphalt content for the mix containing 6% −200 material was selected to be 5.2%. All mixture parameters met the specified requirements at 5.2% asphalt.

The optimum asphalt content for the mix containing 8% −200 material was selected to be 5.0%. All mixture parameters met the specified requirements at 5.0% asphalt. Obviously, these criteria are not sufficient in themselves since Corps of Engineers criteria limits the amount of −200 material to 6%.

The optimum asphalt content for the mix containing 10% −200 material was selected to be 4.8%. All mixture parameters met the specified requirements at 4.8% asphalt; however, the amount of −200 material exceeded the 6% allowed.

The optimum asphalt content for the mix containing 12% −200 material was selected to be 4.6%. All mixture parameters met the specified requirements at 4.6% asphalt, but the amount of −200 material exceeded the 6% allowed.

The addition of −200 material for the mixtures evaluated resulted in a lower optimum asphalt content, higher stability at optimum asphalt content, and an increase in sensitivity to a change in asphalt content. Some filler is needed to obtain the required stability, but excessive filler can result in unsatisfactory mixes. For the mixtures evaluated, it appears that the optimum amount of −200 material was between 3 and 6%.

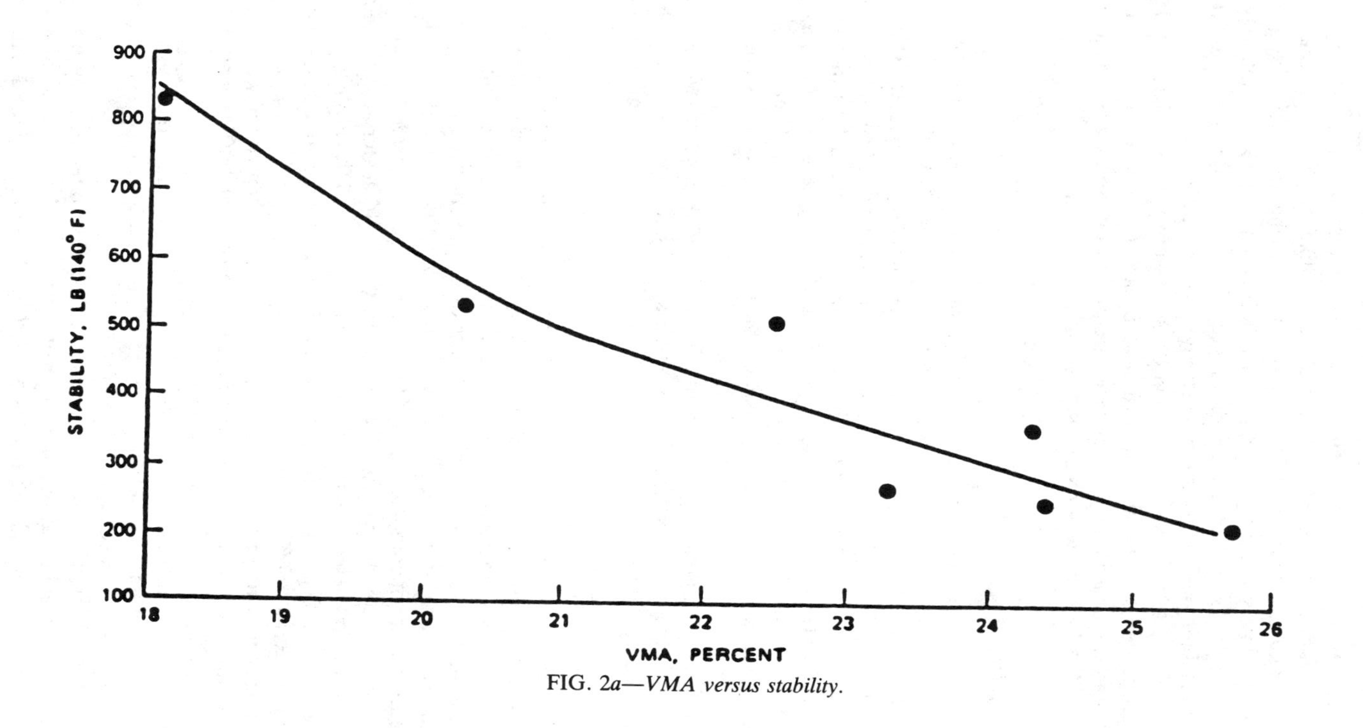

FIG. 2*a*—*VMA versus stability.*

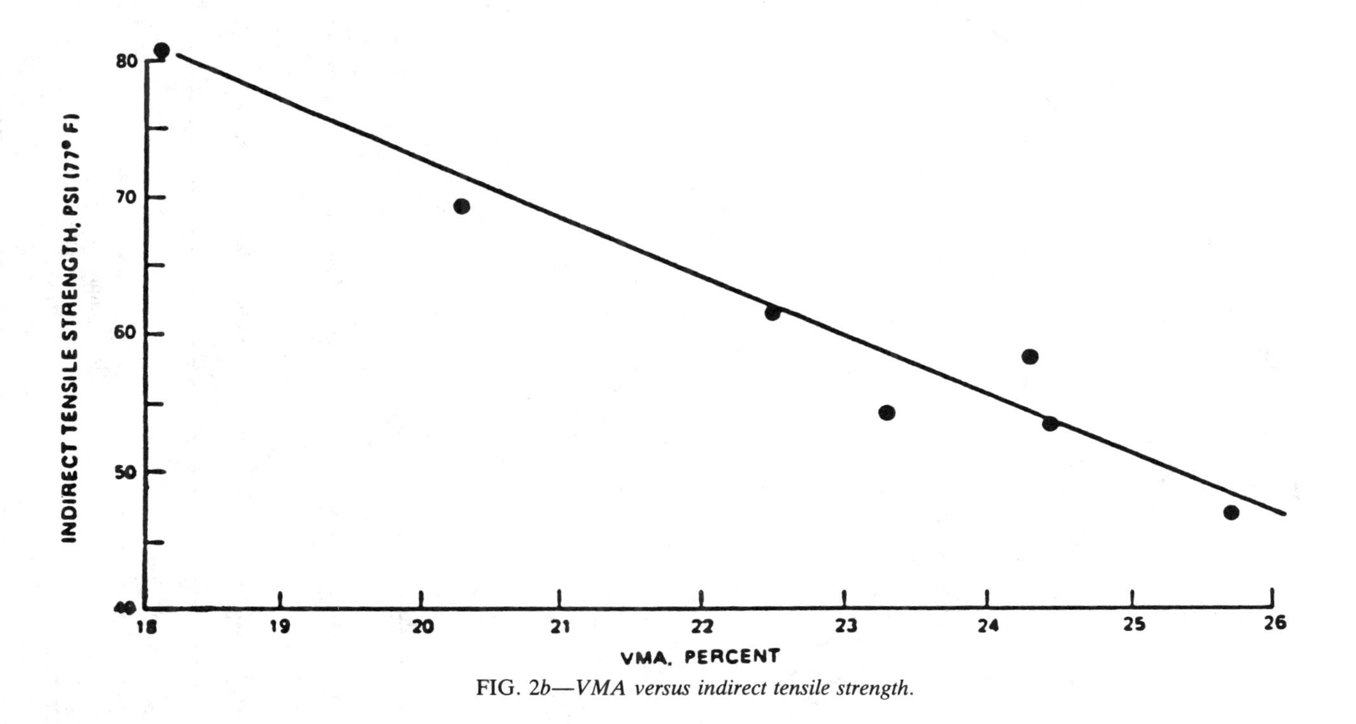

FIG. 2*b*—*VMA versus indirect tensile strength.*

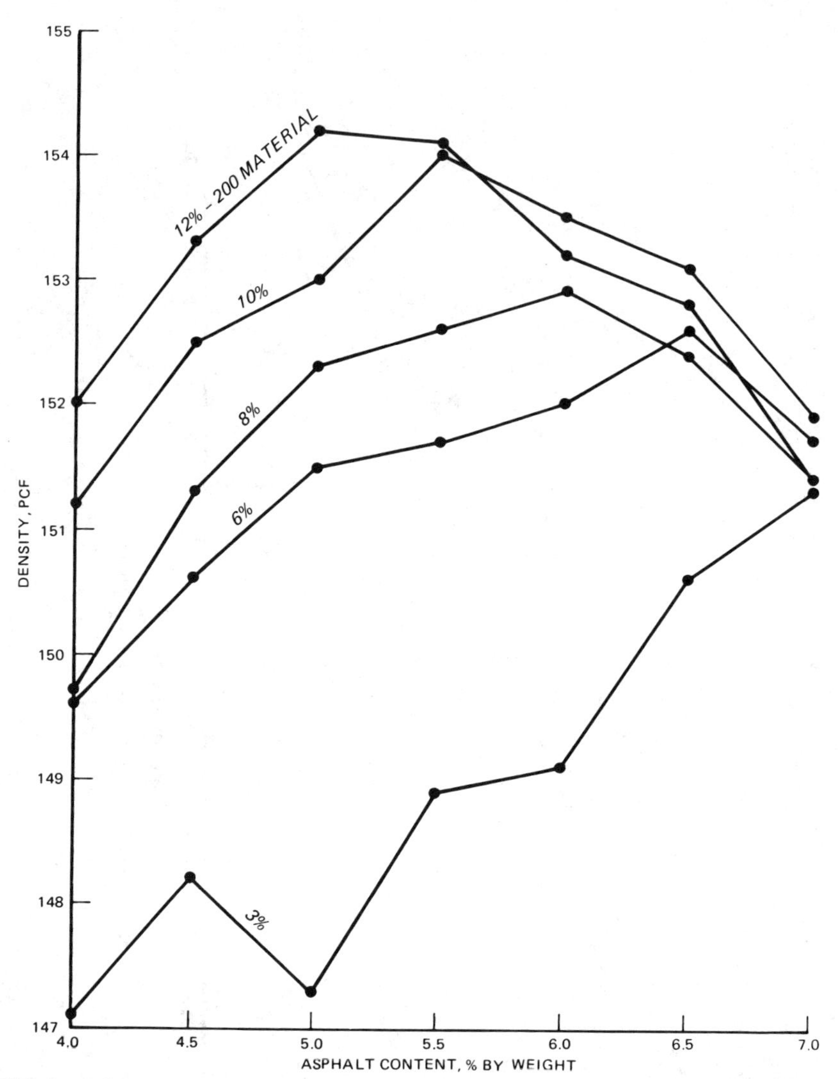

FIG. 3—*Relationship between asphalt content, density, and amount of −200 material in mix.*

TABLE 4—*Comparison of Corps of Engineers criteria for pavements with laboratory-determined values.*

Parameter	Corps Criteria	Laboratory Value
Stability, lb[a]	1800 (minimum)	1620
Flow, 0.01 in.[b]	16 (maximum)	10
Voids in total mix, %	3–5	3.2
Voids filled with asphalt, %	70–80	83

[a] Lb × 0.45 = kg.
[b] In. × 25.4 = mm.

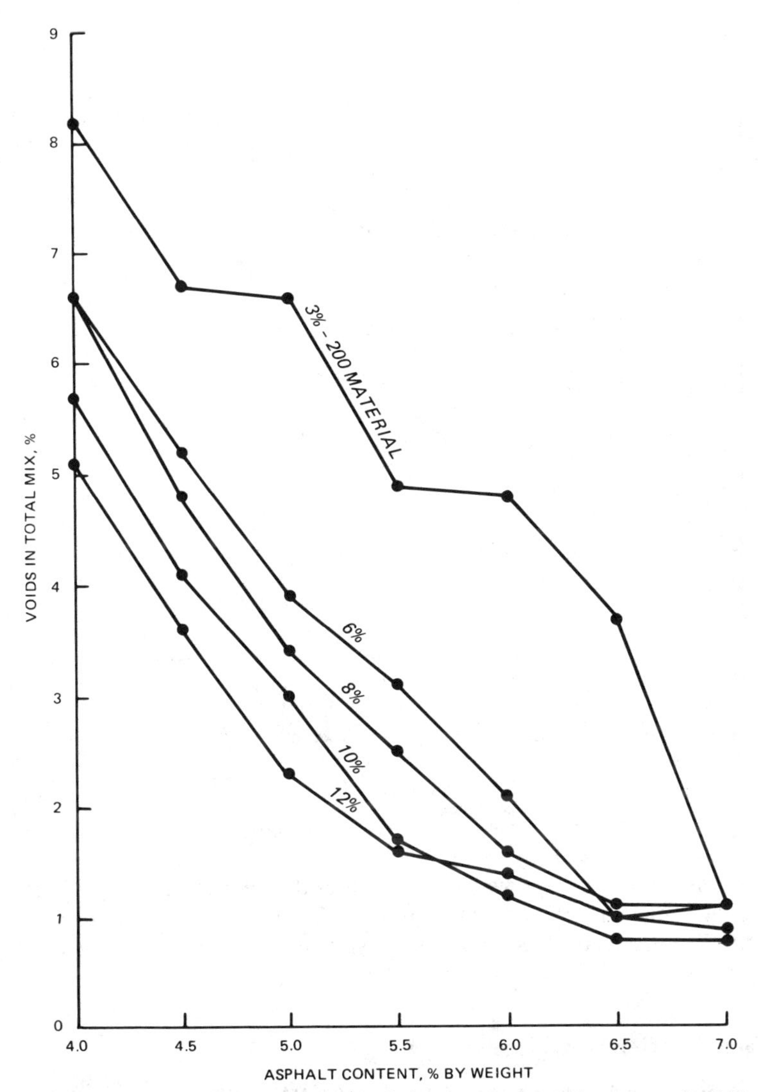

FIG. 4—*Relationship between asphalt content, voids in total mix, and amount of −200 material in mix.*

Effects of Filler on Field Compaction

The amount of mineral filler affects the workability of asphalt concrete. Control charts for an asphalt concrete construction project showed that the amount of filler varied significantly as the project progressed (Fig. 8*a*). A control chart for the laboratory density (Fig. 8*b*) showed an increase in density with increase in the amount of filler. The field density expressed as a percent of the laboratory density showed an inverse relationship between the amount of filler and the field density (Fig. 8*c*). As this project progressed, the effect that the filler was having on the field density was recognized. The contractor was made

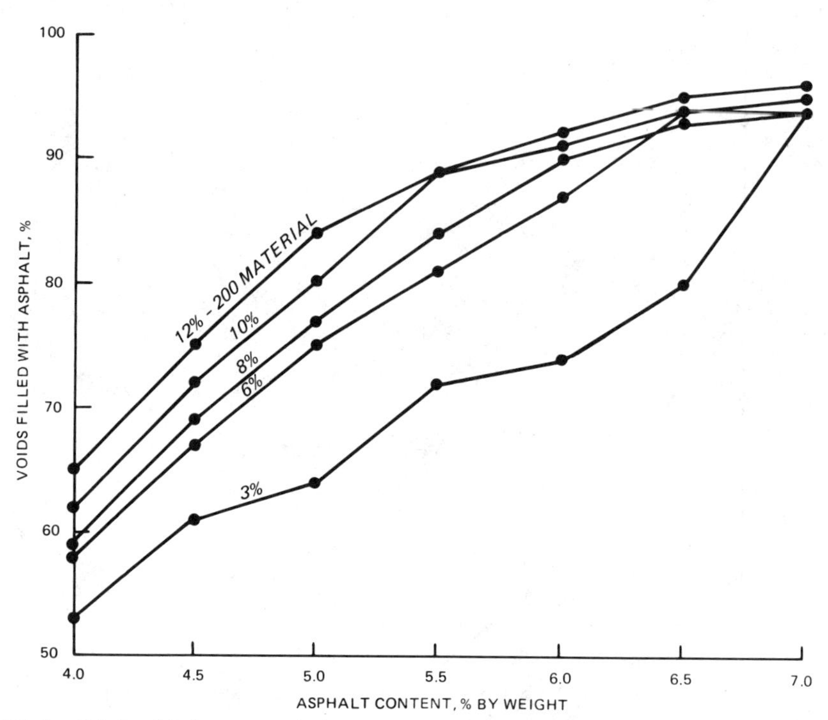

FIG. 5—*Relationship between asphalt content, voids filled with asphalt, and amount of −200 material in mix.*

aware of the problem, and steps were taken to lower the filler content. As indicated in Figs. 8*a* and 8*c*, the reduction in filler content resulted in an increase in field density and compliance with the specification requirements, in spite of the opposite effect indicated by the laboratory tests.

Use of Natural Occurring Aggregates

The Mississippi State Highway Department (MSHD) allows the use of a fine sand aggregate in its asphalt mixtures. The use of this fine sand aggregate is believed to contribute to the rutting problem which has been observed within the state. This fine sand aggregate is actually very nonhomogeneous top soil containing a large amount of low-quality material such as clay and silt, which should not be used in an asphalt mixture. The material is difficult to feed properly through a cold feed. All these factors combine to create variation in the mixture that result in poor performance on the roadway.

Since most natural sands are rounded and often contain a high percentage of undesirable materials, the Corps of Engineers limits the amount of natural sand used in an asphalt mixture to 15% for airfields and 25% for roads. Natural sands are also often rejected due to grading or excessive amounts of undesirable material such as organics or clay balls. Other aggregates used in the mixture are required to be crushed. There is much evidence to show that many pavement problems which have been observed are the direct result of using too much natural sand.

Effect of Maximum Aggregate Size

The maximum aggregate size used in an asphalt concrete mixture greatly affects the mixture performance. Larger aggregate size generally results in better skid resistance, improved mixture stability, and lower optimum asphalt content. In the past five years, the MSHD has experienced rutting problems on a number of primary and interstate routes. During the late 1970s and early 1980s, the MSHD specified open-graded friction courses for primary and interstate routes to improve wet skid resistance of asphalt concrete pavements. Due to the high cost and poor performance of these friction courses, the use was discontinued and the grading requirements for dense graded surface courses was changed to require 100% of the aggregate blend to pass the 0.95-cm (3/8-in.) sieve. Previously it was required that 100% of the aggregate material pass the 1.27 cm (1/2-in.) sieve; however, it was found that crushing to this size resulted in the production of elongated particles that contributed to poor skid properties. As would be expected, the bitumen content had to be

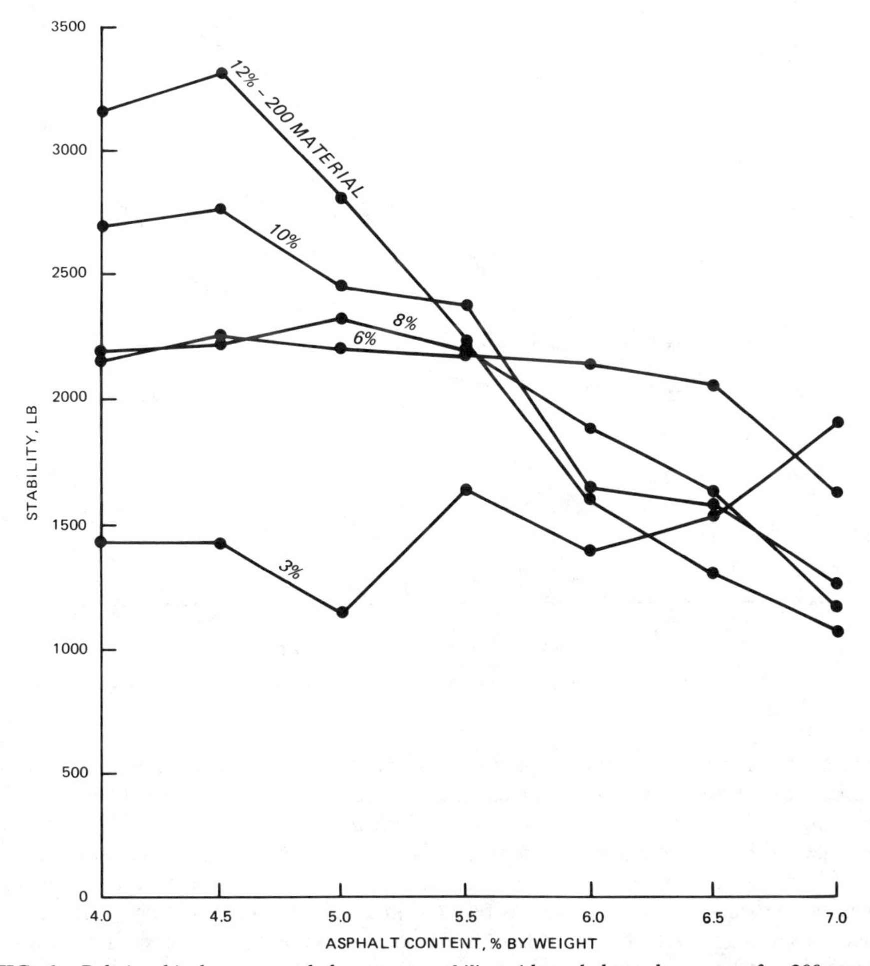

FIG. 6—*Relationship between asphalt content, stability with asphalt, and amount of −200 material in mix.*

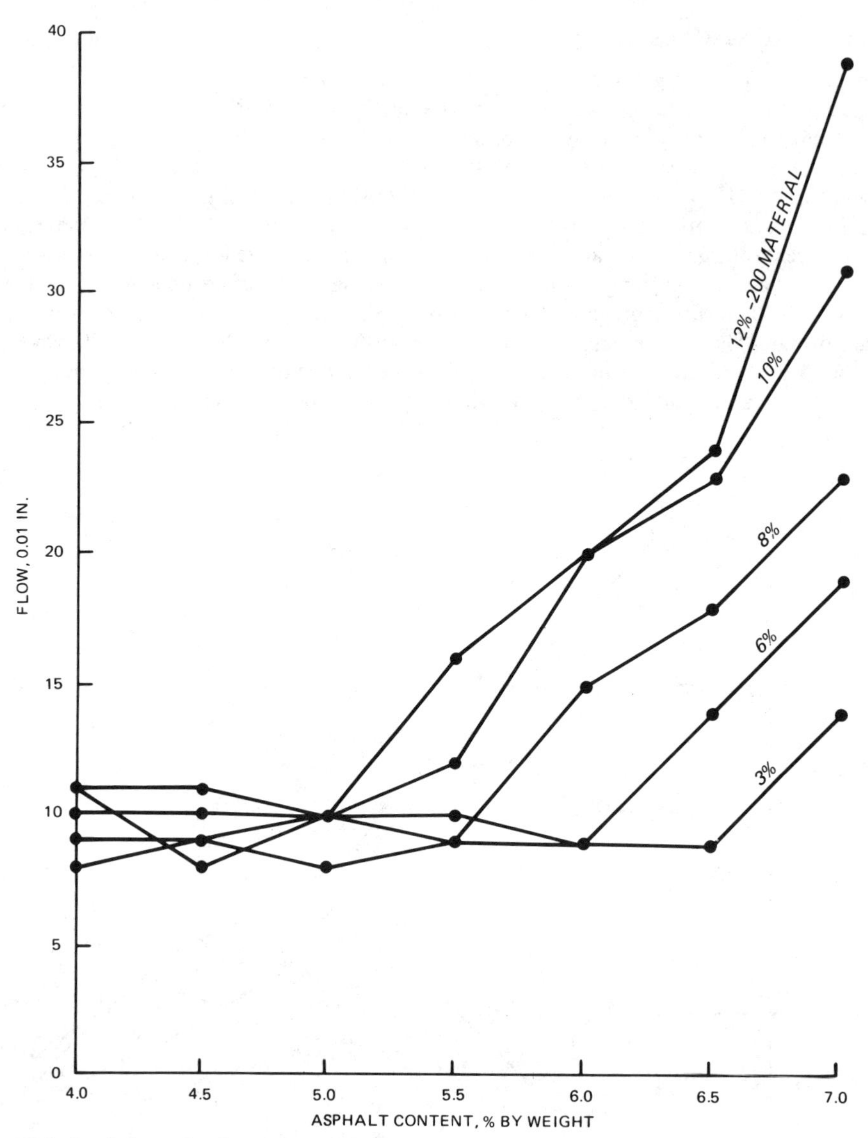

FIG. 7—*Relationship between asphalt content, flow, and amount of −200 material in mix.*

increased when the maximum aggregate size was reduced. More bitumen with a finer gradation increased the number of problems due to instability of the mixture. The mix design now being used is being evaluated for future modifications.

Another problem with aggregates in the MSHD area is the small size of the natural gravel. It is difficult and expensive to produce a 1.27-cm (½-in.) maximum size aggregate gradation that is well crushed. This difficulty has resulted in aggregate being brought in from other states (thus at high cost), the use of finer mixes, or the use of mixtures containing rounded aggregate. Another significant part of the rutting problem in the MSHD is the desire for lower air voids in bituminous pavements. During the 1970s it was discovered that alarmingly high air void contents existed in asphalt concrete pavements and that this was related to shortened service life. The first response was to compact more vigorously. The use of finer mixtures compacted to a low void content in the field requires a considerable increase in asphalt content over that originally used. These changes have resulted in more rutting. The increase in truck tire pressure is also reason for concern.

MSHD Procedure for Evaluating Mixtures [*6*]

The MSHD is now developing a procedure to use the Gyratory Testing Machine (GTM) for testing and evaluating asphalt concrete mixtures during production. The first step in the use of the GTM was to determine the relationship between GTM density and the ultimate roadway density. A ram pressure of 827 kPa (120 psi), with a 1-deg gyratory angle was used,

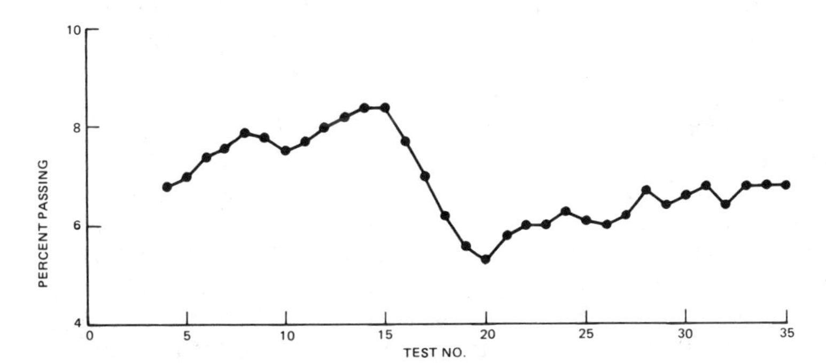

FIG. 8*a*—*Amount of material passing No. 200 sieve.*

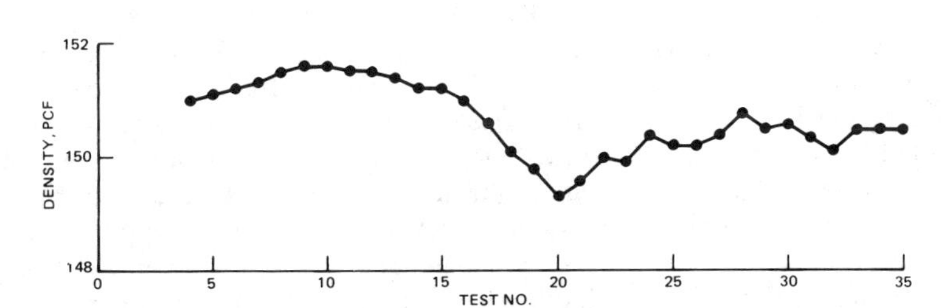

FIG. 8*b*—*Laboratory density (average of last four samples).*

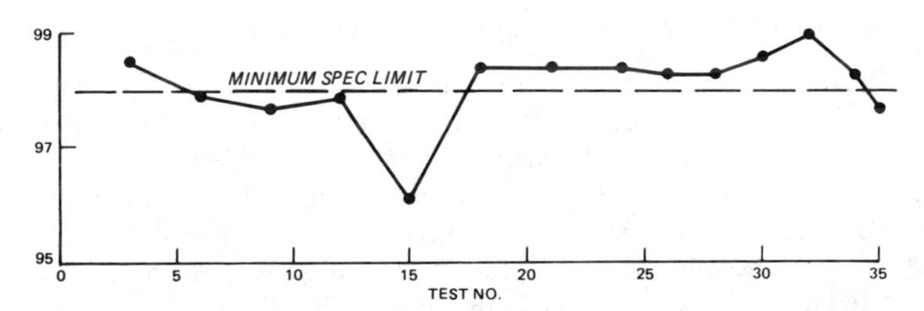

FIG. 8*c*—*Field density, percent of laboratory (average lot density).*

and compaction was generally carried to 200 revolutions. Core samples were obtained from eight locations on interstate and primary four-lane highways. After the in-place density was determined, the cores were heated and recompacted in the GTM at various numbers of revolutions. Measurements of sample height and roller pressure were generally made at 30-revolution intervals along with gyratory shear strain as indicated by the gyrograph. Test results were then plotted and modeled with a parabola by the least squares method. The equation obtained from this procedure is

$$Y = 97.378 + 0.029X - 6.045 \times 10^{-5}X^2$$

where X = GTM revolutions and Y = GTM density as a percentage of roadway density.

This equation indicated that 100% of field density is achieved at approximately 120 revolutions when heating and recompacting roadway samples directly. There is evidence that this correlation is not valid for use in design tests because of the aging effects. The GTM parameters reflect this aging. We believe that, although it is beyond the scope of this paper, a satisfactory correlation will need to be based upon tests on the original or reconstituted mixes instead of the weathered mix from the road.

In the evaluation of existing pavement by recompaction in the GTM, parameters of density, Gyratory Elasto-Plastic Index (GEPI), and roller pressure reflect certain important properties related to aging. The evaluation is enhanced if the bitumen is extracted and a "reconstituted" mix is made using the same amount of fresh bitumen. This is illustrated in Fig. 9, which compares the results for a limestone mix and a chert gravel mix. The samples beginning with "L" are the limestone mixes; the samples beginning with "J" are the gravel mixes. The subscript "O" indicates the original mix, while "R" indicates the reconstituted mix. It is noted for the limestone mix that two of the three GTM parameters (density and GEPI) show minimal difference between the original and the reconstituted mixes. This is interpreted to mean this pavement has experienced minimal aging after four years in service. The chert gravel mix, on the other hand, shows quite large differences for these same two parameters after five years of service. These differences indicate that significant changes have occurred due to aging. The reduction in the gyrograph band width (negative slope to the GEPI curve) and the reduction in density apparently reflect a serious reduction in the quality and bonding of the bitumen. Apparently, the bond between aggregate and bitumen had already been lost because of the hydrophylic (water-loving) quality of the chert gravel and because stripping had occurred under the action of traffic. As kneading takes place in the GTM, the effects of stripping are immediately evident from the negative slope of the GEPI curve. This is evidently caused by a buildup of internal friction as the aggregate particles are directly exposed to each other because the oxidized asphalt loses its bond and no longer furnishes lubrication and cohesion. Also, without the lubrication normally provided by the asphalt, lower densities are achieved during GTM recompaction of the field samples.

MSHD Evaluation of Bituminous Mixes at Hot-mix Asphalt Plants

Field trips were made to seven sites in five of the six districts in Mississippi to evaluate bituminous mixes being produced for active construction projects. These field trips were conducted during the period August 1984 through November 1985.

A mobile field unit was prepared by installing a GTM in a Chevrolet van, Model 30. The general procedure at each site was to run, initially, three or four sets of tests to determine the characteristics of the mix and the degree of variability. Approximately 8 kg (15 lb) of mix was collected for each set of tests, and this mix was placed in a portable oven to maintain the temperature at about 79°C (175°F). From this amount of mix, three specimens were

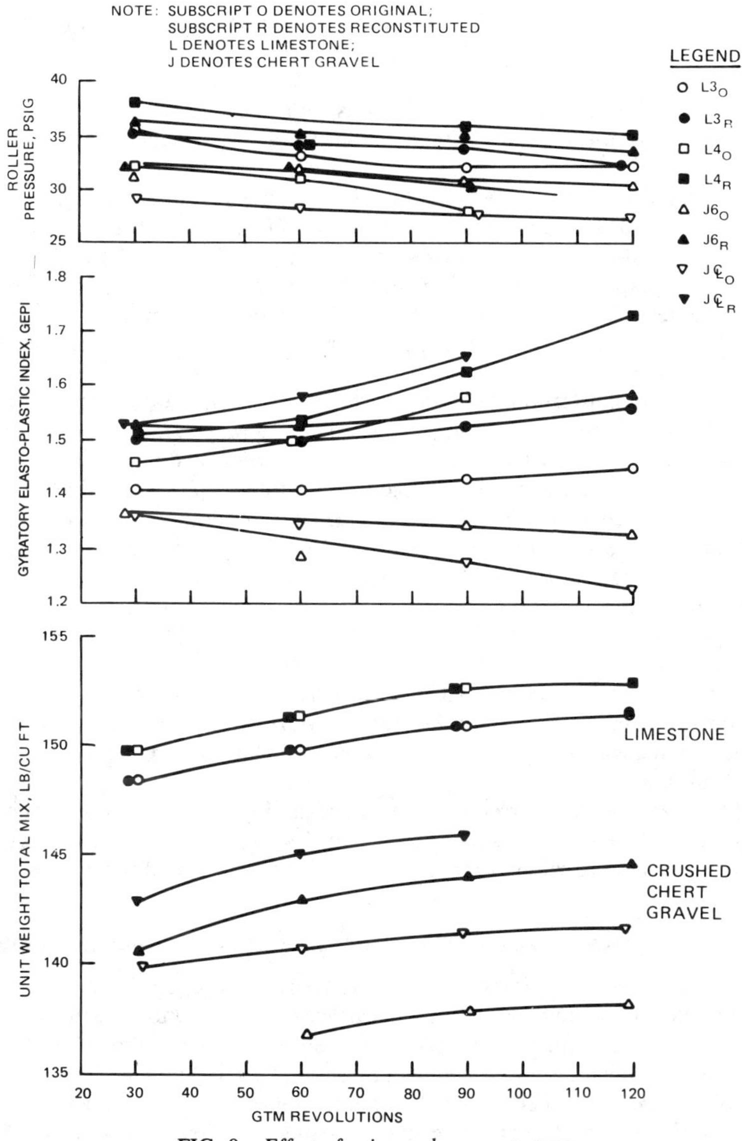

FIG. 9—*Effect of aging and aggregate type.*

usually taken and tested in the GTM, two specimens at a ram pressure of 827 kPa (120 psi) and one specimen at a ram pressure of 1379 kPa (200 psi). A gyratory angle of 1 deg was used for all tests. When possible, the sample was collected from the same truck that was sampled by the field lab technician in order that the bitumen content and the gradation of the sample would be available. Where the field lab was equipped with a nuclear asphalt content gage, it was used to determine the bitumen content for each set of tests.

The brief analysis of the test data was performed immediately after each test. Based on the results of the core analysis, it was decided that an optimum mix for heavily trafficked routes, like those represented in the core study, should have a GSI meeting the tentative criteria

$$\text{GSI} = 1.0 \text{ at 90 revolutions (to avoid rich mix)}$$
$$\text{GSI} > 1.0 \text{ at 150 revolutions (to avoid lean mix)}$$

The gyratory angle was set at 1 deg and the ram pressure was set at 827 kPa (120 psi).

A simple index was identified that is very helpful in explaining the behavior of mixes. The index is the sum of the percentages of bitumen content and −200 sieve aggregate fraction. Also, a graphical illustration of the GTM test results was used to present a good summary of the tests. The vertical axis of the graph is divided into three parts where density, Gyratory Stability Index (GSI), and upper roller pressure are plotted versus GTM revolutions, which is the horizontal axis.

For the purpose of illustration, test results from one field trip are presented here. A drum plant located at McComb, Mississippi, was visited during May 1985. Test results are shown in Fig. 10. As the legend indicates, the bitumen content remained fairly constant, while the −200 fraction changed considerably. As the index (bitumen plus <200) increased, higher densities were achieved. Correspondingly, the GSI climbed, indicating overfilled voids. The magnitude of the roller pressures and the slopes of roller pressure curves indicated moderately reduced strength of the mix as the index increases.

Results similar to these were documented at plants in other parts of the state. This research indicates two overriding concerns:

1. The mix produced in the asphalt plant usually has vastly different GTM properties from the mix designed in the Central laboratory. More rigorous testing needs to be done in the field to initiate plant production and to ensure that the mix being produced meets the design criteria.
2. The fine sand aggregate currently being used is causing problems that are compromising the quality of the pavements. It should be replaced with a mineral filler that will provide the necessary particle size without the inherent problems associated with the fine sand.

Critical Bitumen Content: Function of Aggregate Properties and Compaction Efforts

Experience has shown that bituminous pavement performance can be conveniently related to the voids content only when the testing of the aggregate and the mix, and the interpretation of the results, are appropriately done by experienced pavement design engineers. The relationship is quite complex and fraught with pitfalls! To quote the late Bruce G. Marshall [5], "no limits can be established for VMA, for universal application, because of the versatile application of bituminous materials to many types and gradations of aggregate." Some test results taken from Marshall [5] are shown in Fig. 11, where percent VMA (voids in the mineral aggregate) versus percent coarse aggregate is shown for specimens at Marshall 50 blow optimum. The aggregate gradations used for these tests are shown in the accompanying tabulation in Fig. 11 (see also Table 5). Marshall [5] cites these data, along with similar results for VMA versus coarseness of sand and VMA versus percent passing No. 200 screen to demonstrate clearly that the voids in the mineral aggregate at the peak of the compaction curve is a variable function of gradation. Marshall [5] considered the peak of the unit weight total mix curve to be the optimum for the mix and points out that consequently no one set of voids criteria can be established for universal application. (The reader is reminded that the original Marshall Procedure and criteria are not the same as those of the Corps of Engineers.)

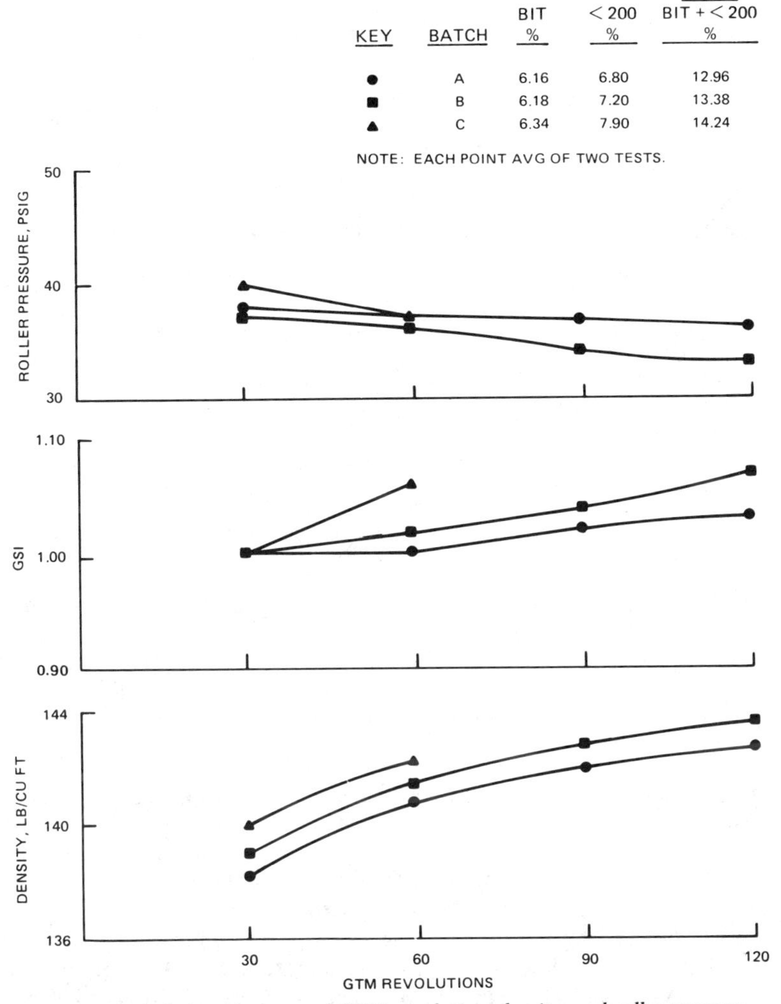

FIG. 10—*Relationship between GTM revolution, density, and roller pressure.*

Figures 12 and 13 are presented to illustrate further the effect of the gradation of the aggregate on the voids total mix and voids filled at the peak of compaction curve [*1*]. Compaction results for these mixes are shown on Fig. 13, where it is evident that the critical void contents (at peaks of compaction curves) are significantly different for these mixes even though these gradations would fall within many gradation specification limits. Since the compaction test characterizes the fundamental physical properties of the mix, it is evident that this *critical void content is a variable,* as is the critical bitumen content, Mix 1 being characterized by 1.5% voids at 5.0% bitumen while Mix 2 is characterized by 5.0% voids at 4.5% bitumen, with the rather surprising evidence that the denser graded mix has a higher compaction optimum bitumen and the "hump" graded mix. Although it may appear that these two mixes represent extremes, in actuality wide variations in the voids content at optimum is more the rule than the exception. This is exemplified by results extracted from

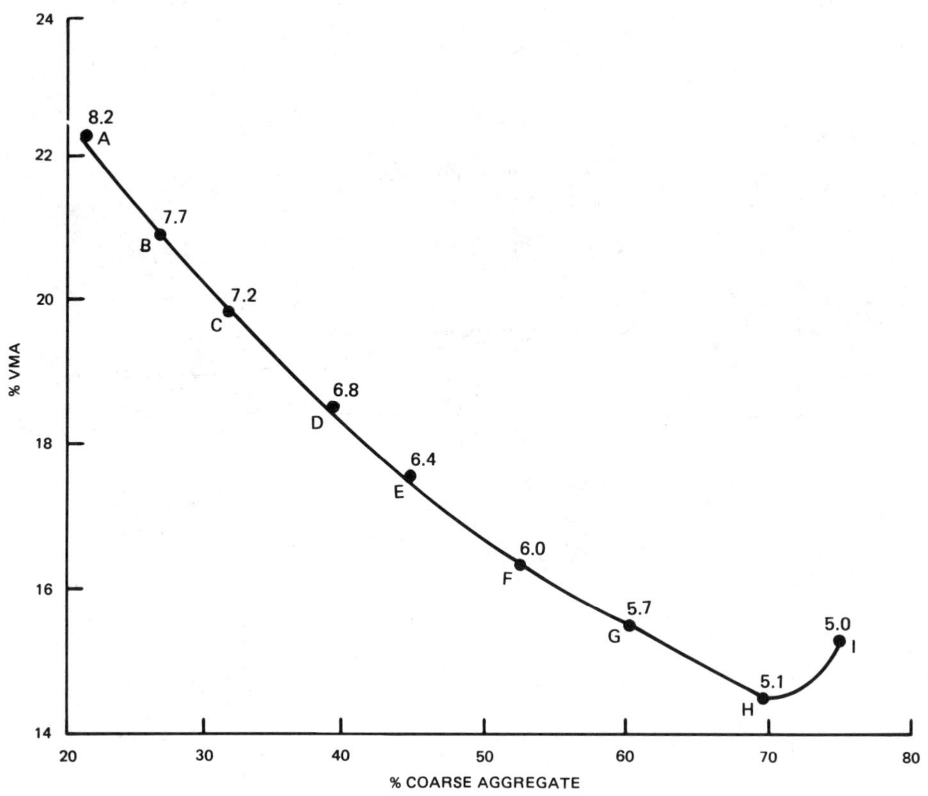

FIG. 11—*Percent coarse aggregate VMA versus percent coarse aggregate* (*after Marshall*).

reports by Philip J. Arena of the Louisiana Department of Highways [*8*] and E. R. Brown of the U.S. Army Corps of Engineers Waterway Experiment Station [*4*].

First data will be cited from Arena [*8*]. Figure 14 shows the gradation of five different mixes for which the voids for the compaction test results are given in Table 6. The compaction tests included both mechanical and manual Marshall hammers (50 blows and 75 blows) and

TABLE 5—*Aggregate gradation for VMA determination.*

Gradation Number	¾-in. (1.9 cm)	½-in. (1.27 cm)	⅜-in. (0.95 cm)	No. 4	No. 10	No. 40	No. 80	No. 200
A	...	100	97.5	91.0	80.0	51.0	32.0	10.0
B	100	98.0	94.9	86.0	74.0	42.5	25.5	8.5
C	100	96.0	91.5	81.0	68.0	39.0	23.4	7.8
D	100	93.8	88.9	76.2	61.5	35.2	21.0	7.0
E	100	91.0	84.3	69.0	54.0	31.0	18.6	6.2
F	100	89.0	80.8	62.0	47.0	27.0	16.2	5.4
G	100	86.0	77.0	55.0	40.0	22.0	12.0	4.0
H	100	76.0	58.0	46.0	35.0	20.0	12.0	4.0
I	100	76.0	58.0	40.0	30.0	17.2	10.3	3.4
J	100	76.0	58.0	35.0	24.0	10.0	6.0	3.0

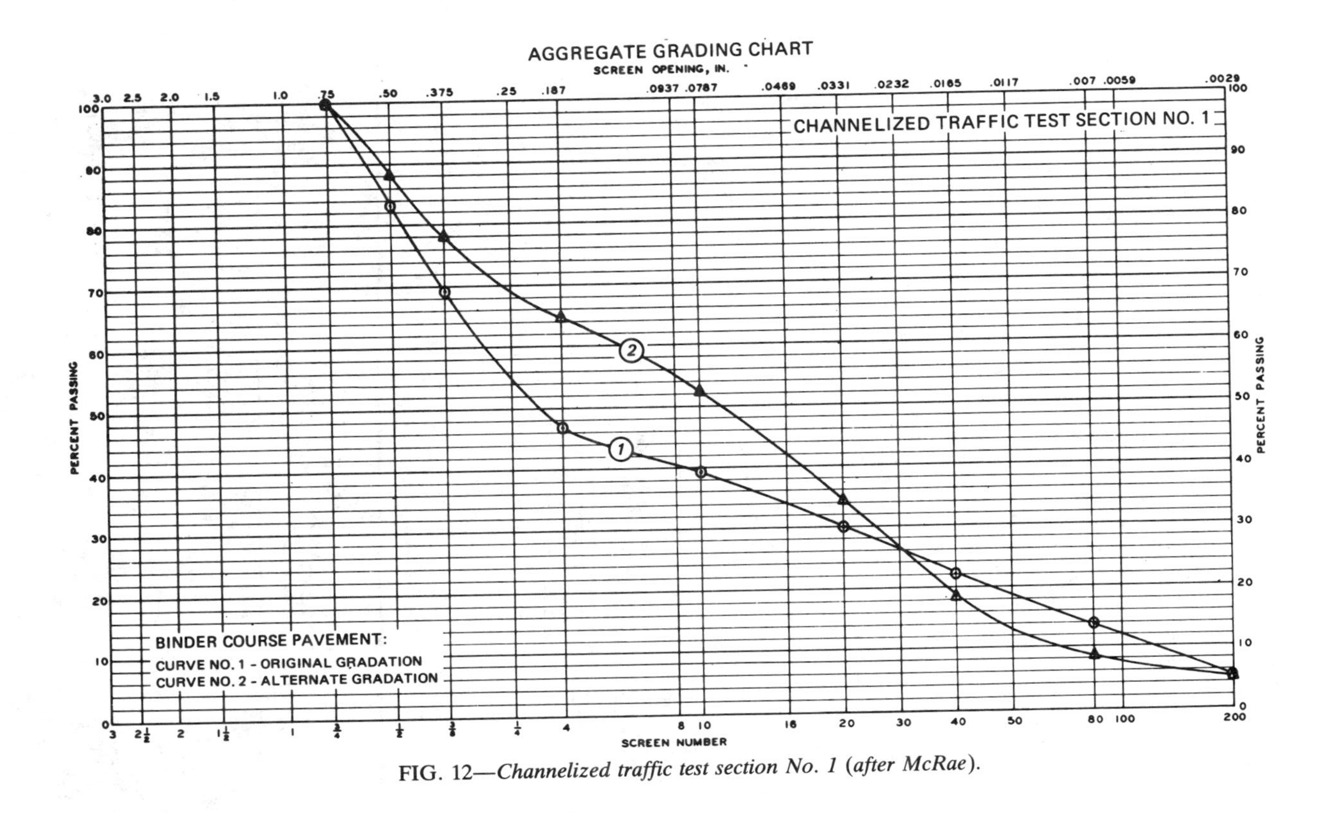

FIG. 12—*Channelized traffic test section No. 1 (after McRae).*

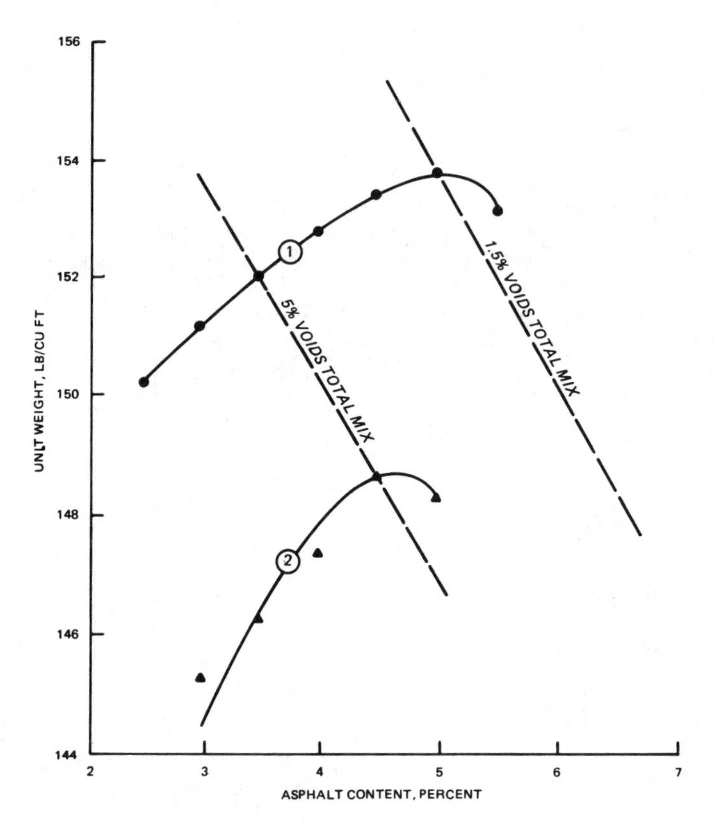

FIG. 13—*Channelized traffic test section No. 1 (after McRae).*

TABLE 6—*Percent void total mix.*

Compaction	Mix 1 Optimum 4.8%	Mix 2 Optimum 5.2%	Mix 3 Optimum 5.2%	Mix 4 Optimum 4.6%	Mix 5 Optimum 4.5%	Variation in Critical Voids with Gradation
50 blow (hand)	6.0	7.8	5.8	3.0	4.7	3.0–7.8
50 blow (mechanical)	7.0	8.8	7.4	3.3	6.2	3.3–8.8
75 blow (hand)	5.5	6.0	5.7	2.8	3.7	2.8–6.0
75 blow (mechanical)	6.3	8.5	5.8	2.9	4.2	2.9–8.5
689 kPa (100 psi), 1 deg, 60 revolutions	4.8	4.5	5.4	2.3	4.0	2.3–5.4
Variation in critical voids with compaction	4.8–7.0	4.5–8.8	5.4–7.4	2.3–3.3	3.7–6.2	

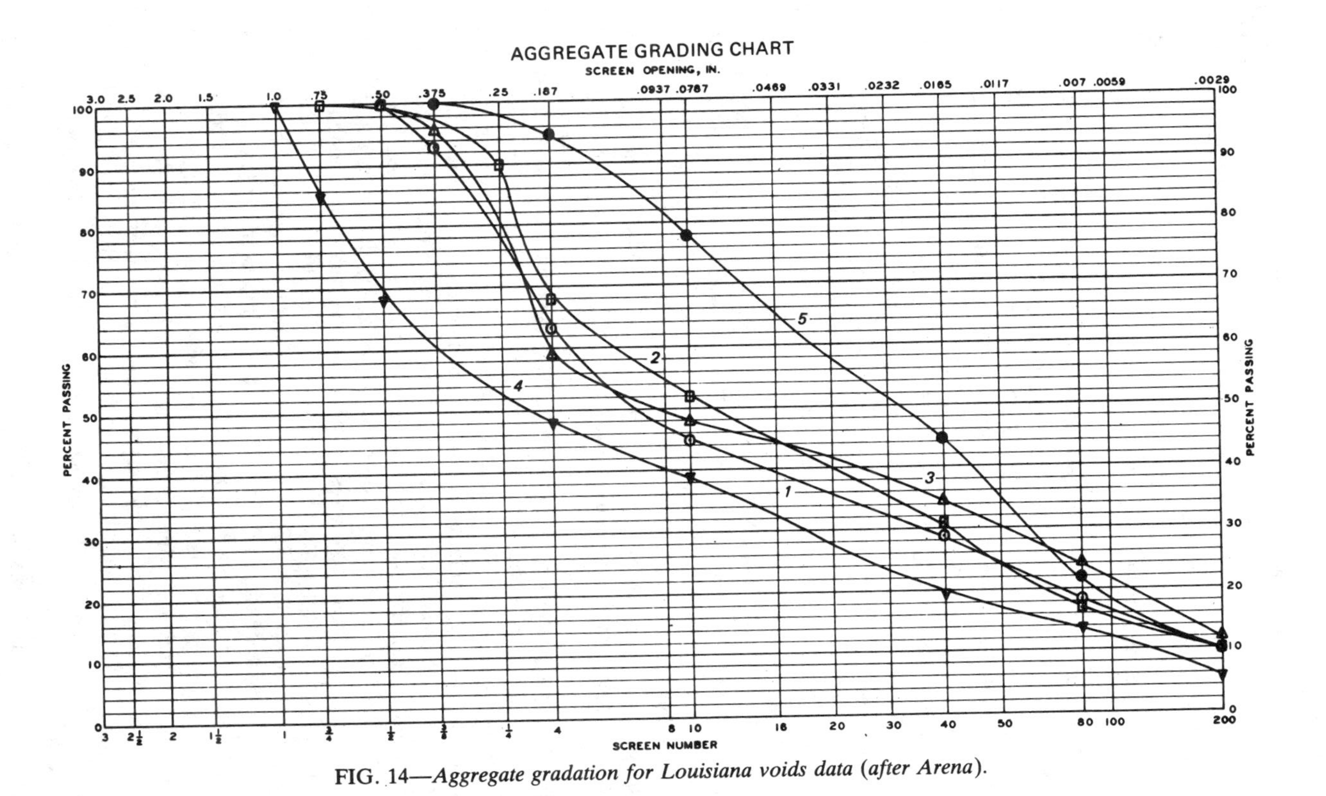

FIG. 14—*Aggregate gradation for Louisiana voids data (after Arena).*

compaction in the Gyratory Testing Machine (GTM) using 689 kPa (100 psi), 1 deg, 60 revolutions.

It will be readily apparent to the reader that Marshall [5] was "on target" 37 years ago when he said: "The compactive effort applied in the production of test specimens has a direct bearing on all the physical properties. . . . The compactive effort applied to specimens for establishing the asphalt content and testing purposes must produce density equivalent to that which will be ultimately developed under traffic."

The Louisiana Department of Highways, along with many others, recognized this need. Quoting from the introduction of Arena's paper [*8*]:

> During the past several years the Louisiana Department of Highways, through necessity, had to increase the intensity of the pneumatic rollers for compacting asphaltic concrete pavements. This, consequently, made it necessary to increase the compactive effort of the laboratory design of asphalt concrete mixtures. The need for this increase had become critical due to the excessive rutting and shoving observed on asphalt concrete pavements after being subjected to traffic.
>
> In an attempt to remedy this problem, high intensity pneumatic rollers capable of exerting contact pressures of up to 90 psi were incorporated into the specifications. To supplement this, the laboratory design compactive efforts were increased from 50 blows to 75 blows on both sides of a 101.2 mm (4-in.) diameter specimen using a standard Marshall impact hammer.
>
> Although these modifications have shown a vast improvement in asphaltic concrete pavements in Louisiana, it again appears that an additional increase in design compactive effort is essential in obtaining maximum design life, due to the rapid increase of traffic volume encountered on the highways.
>
> One of the objectives of this study then, is to establish an adequate laboratory compactive effort for design of asphaltic concrete by use of the Gyratory Compactor.

Figure 15 shows the correlation between GTM compaction and the roadway at 6, 15, and 36 months [*8*]. It appears that the ultimate density for this mix under 689-kPa (100 psi) compaction pressure is attained at the asymptote of the curve at about 70 gyrations. The number of gyrations or revolutions required to reach the ultimate roadway density is expected to vary with the compactability of the mix, but for any given design stress the ultimate density should be closely associatd with the asymptote of the density versus revolutions curve. Since the Louisiana correlation curve [*8*] (Fig. 15) flattens out at 3.5% voids, it is quite likely that the bitumen content is somewhat on the "rich" side and that ultimate density has been attained a bit prematurely. That is, the optimum for this traffic and this GTM effort is probably lower than 5.8%, in which case the pavement would have taken longer to reach the ultimate density and the number of revolutions in the GTM would be greater. More will be said about compaction presently, but first, some data will be presented on the variation in voids of pavements experiencing varying degrees of rutting.

The data immediately following are extracted from Brown's paper [*4*]. This paper presents the results of the evaluation of 23 primary taxiway pavements at seven different Air Training Command (ATC) airfields, subjected to T-38 aircraft loadings. The T-38 aircraft usually operates with a tire pressure of 1654 kPa (240 psi) and a wheel load of 2475 to 2588 kg (5500 to 5750 lb).

Figure 16 shows the Corps of Engineers specification band for the gradation of these pavement mixes. The results presented here are for the surface course or top layer of pavement, and only the deviations from the specification grading band are plotted.

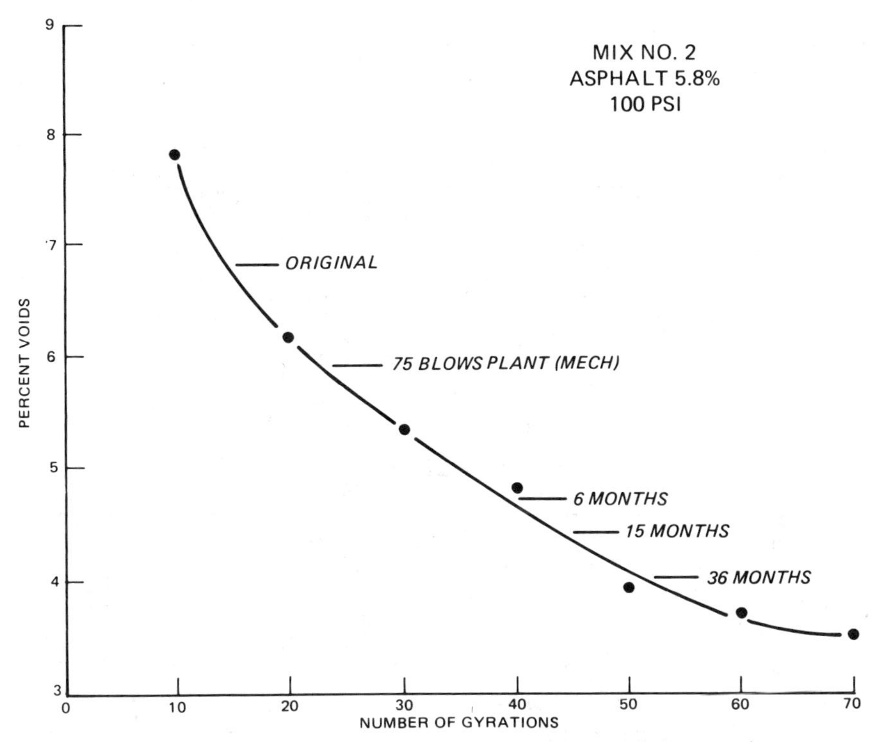

FIG. 15—*Evaluation of gyratory testing machine (after Arena).*

Each pavement was evaluated with regard to degree of rutting as indicated in Table 7, which also shows the results of recompaction in the GTM. Core samples of pavement were taken for density and voids determination, results of which are tabulated in Table 8. It is evident from the data in Table 8 that the criteria of 3 to 5% voids total mix and 70 to 80% voids filled are "out of focus" as discerners of quality for this pavement; in fact, these results nullify the efficacy of these voids criteria as precision indicators of the flushing or rutting phenomenon for these pavements. Regardless of whether the pavement experienced rutting or not, the scatter in the VMA determinations is relatively narrow and these 14 mixes at seven airfields approach an average of 14%.

The results of tests with the GTM shown in Table 6 and Fig. 17 show clearly that the compaction phenomenon in conjunction with the GTM "Gyrograph" was the most reliable indicator of the maximum permissible bitumen content for these pavements. The compaction effort used in the Brown paper [*4*] was 1379 kPa (200 psi), 1 deg, 30 revolutions. Quoting Brown [*4*], "As the 30 revolutions are being applied to the sample, the sample densifies to a point at which, due to low void content the mix becomes unstable. . . . When this happens, shear failure takes place. . . . Hence, the deformation (apparent compaction) being obtained under the load causes an upheaval of the material adjacent to the load. . . . This simulates what happens when rutting occurs in the field." Apparently, the GTM is a good mechanical analog of the pavement that it is able to "take account" of the multitude of variables, including not only the degree of compaction and the gradation, as already discussed, but

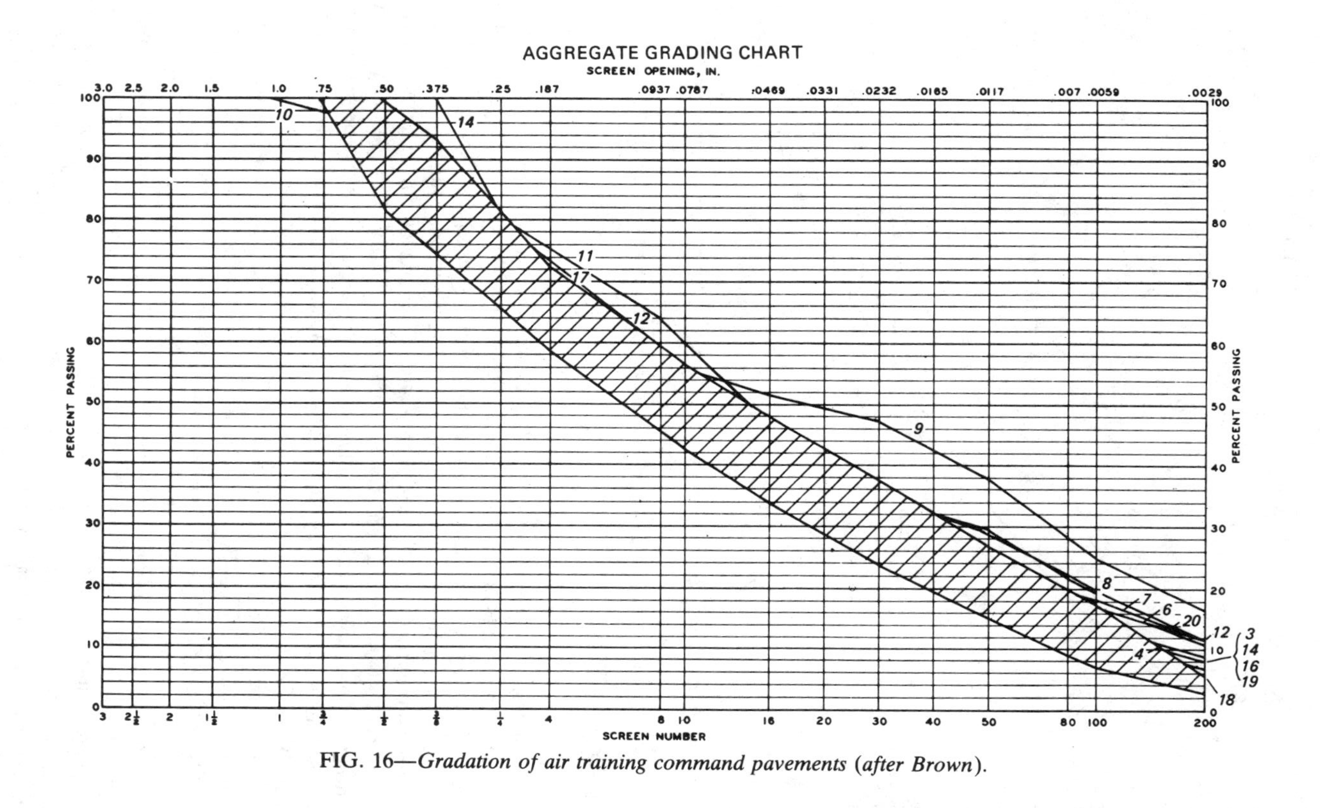

FIG. 16—*Gradation of air training command pavements (after Brown).*

TABLE 7—*Test results for surface course (top layer) field cores.*

Air Force Base	Taxiway	Specimen No.	Voids: Aggregate Only	Voids: Total Mix	Voids: Filled	Degree of Rutting[a]	Gyratory Flushing: Yes	Gyratory Flushing: No
Craig	2A	1	20	10	53	slight		x
Craig	3	2	15	8	58	slight		x
Craig	3A	3	16	6	73	severe	x	
Craig	3D	4	12	6	64	none		x
Craig	7	5	16	9	56	slight		x
Laughlin	1	6	12	5	74	slight	x	
Laughlin	1A	7	14	7	60	severe	x	
Laughlin	1B	8	13	5	74	slight	x	
Moody	4	9	14	6	68	slight		x
Moody	10	10	13	6	69	slight		x
Reese	7	11	13	4	78	slight	x	
Reese	8	12	15	6	71	severe	x	
Reese	10	13	18	10	58	slight	x	
Vance	1	14	13	6	66	slight	x	
Vance	3	15	14	7	58	none		x
Vance	14A	16	12	5	69	none	x	
Vance	14B	17	14	7	60	slight		x
Vance	15A	18	14	7	57	severe		x[b]
Vance	15B	19	16	10	48	none		x
Webb	4	20	11	3	81	severe	x	
Williams	3	21	16	9	58	none		x
Williams	7	22	15	8	61	none		x
Williams	8	23	16	8	59	none		x

[a] Definition of rutting: slight = any ruts up to ½ in. (1.27 cm) deep; severe = any ruts over ½ in. (1.27 cm) deep.

[b] GTM Test on the underlying second layer showed flushing which accounts for severe rutting even though the surface layer did not indicate flushing in the GTM.

TABLE 8—*Average test results for surface course (top layer)—field cores.*

Air Force Base	Average Percent Voids: Voids in Mineral Aggregate (VMA), Degree of Rutting: None	Slight	Severe	Voids in Total Mix, Degree of Rutting: None	Slight	Severe	Voids Filled, Degree of Rutting: None	Slight	Severe
Craig	12	17	16	6	9	6	64	56	73
Laughlin	...	13	14	...	5	7	...	74	60
Moody	...	14	...	...	6	...	...	69	...
Reese	...	16	15	...	7	6	...	68	71
Vance	14	14	14	7	7	7	58	63	57
Webb	...	...	11	...	...	3	...	...	81
Williams	16	...	...	8	...	...	59	...	...
Range	12–16	13–17	11–16	6–8	5–9	3–7	58–64	63–74	71–81
Avg	14	15	14	7	7	6	60	66	68

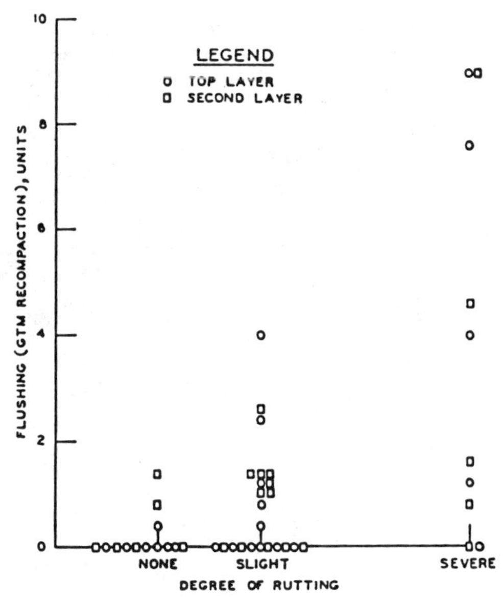

FIG. 17—*Flushing resulting from GTM recompaction versus degree of rutting (after Brown).*

also other aggregate properties such as the porosity or absorption, which complicate the measurement of aggregate specific gravity used in calculating voids. The term "low void content" in the above quote from Brown should not be construed to mean a specific voids content such as 3% or less, but should be interpreted to mean the lowest acceptable void content for the particular mix. This critical voids content is the lowest acceptable for each mix so tested, but it may vary over a considerable range from mix to mix, as has been demonstrated by these data. It is emphasized that the vertical stress used in the GTM should represent the traffic loading and that the ultimate density should represent the pavement density after traffic. Density determination for the pavements listed in Table 6 showed an overall average of 99.3% of the density obtained the GTM, showing a good correlation between GTM density and in-place pavement density.

In the light of current knowledge, not only must the ultimate density be produced in defining the optimum condition but it should be done by a kneading process employing the anticipated pavement design stress [*1*]. This is because the internal particle structural orientation and arrangement is a function of the compaction process and the structural arrangement of the compacted aggregate has a profound effect on the stress-strain properties of the compacted mix [*1,7*]. This requires an acceptable mechanical analog of the pavement in the laboratory test. Ironically, Marshall's basic concept of requiring the ultimate pavement density for the laboratory test still holds, while his method of test has not been able to keep abreast of the modern requirements for density and simulation of the stress-strain properties of the prototype.

McRae [*1*] has advanced the hypothesis that the optimum condition is represented by a statistical equilibrium between the applied load (anticipated design stress) and the body forces of the particles, concurrently at a minimum internal energy situation and close to but slightly greater than the bitumen content at which the maximum unit weight of the aggregate only occurs. It is believed that the GTM employing kneading compaction under the anticipated design stress and employing the gyrograph to sense plasticity identifies this condition for all practical purposes.

The GTM mechanical analog employs kneading compaction at the anticipated design stress, establishes the maximum value for unit weight aggregate only (minimum VTM), and employs a plasticity sensing element (Gyrograph) to indicate, with precision, the bitumen content at which excess plasticity develops. If at the optimum thus established the mix fails to meet the strength requirement or the durability requirement or both (as determined by direct physical tests such as immersion/compression, freeze/thaw, porosity), then adjustments must be made in the gradation or aggregate type or both and a new optimum found by repeating GTM compaction in conjunction with the plasticity-sensing Gyrograph. It is recognized that the percent voids at the optimum is affected strongly by aggregate gradation and porosity as well as by the compaction effort, and that there is an optimum voids condition which varies and which characterizes each mix, just as there is an optimum bitumen content which varies and characterizes each mix. When the anticipated design stress is employed in the GTM, widening of the Gyrograph identifies this condition with due regard for the voids in the mix, as well as the plasticity and shear properties, thus avoiding the necessity of the empirical correlation with the voids.

The shear strength, at the optimum, as measured by the GTM should reflect an acceptable factor of safety against the theoretical maximum induced shear. For convenience, and particularly for monitoring plant production, the GTM roller pressure can be empirically related to gyratory shear which can be rationally related to the maximum induced shear.

Summary and Conclusions

Since aggregate composes approximately 95% of asphalt concrete mixtures by weight, its quality is important. It is difficult to describe aggregate quality with other than gradation and descriptive words such as hard, angular, clean and durable. At best, the aggregate tests used today are indirect indicators and provide only a general indication of quality.

The Gyratory Testing Machine (GTM) can be used to evaluate the effect of aggregates on the total mixture performance. The GTM is quite effective in monitoring plant production with regard to variations in the aggregates. The GTM can be used during mixture design to compact samples to simulate density obtained under traffic and to establish the maximum permissible bitumen content at this density independently of voids criteria. Experience has shown that the GTM can identify mixtures that perform poorly when subjected to high pressure tire traffic.

Further work is needed, and is in progress, to strengthen the correlation of the GTM (which is a mechanical analog of the field) with in-place pavement performance.

The method that is normally used to select aggregates includes using an aggregate source that has a record of satisfactory performance, specifying an aggregate gradation, and specifying a minimum percentage of crushed faces. This paper has provided some guidance on the effects of aggregate gradation and crushed faces, and on recommended methods for evaluating an asphalt concrete mixture to insure satisfactory performance.

The following conclusions are made concerning effects of aggregate gradation and crushed faces.

1. Increasing the maximum aggregate size generally results in a lower VMA or, conversely, a higher unit weight aggregate only, requiring less asphalt and resulting in a tougher mixture.
2. A small change in type or amount of mineral filler may significantly change the compactability as well as other characteristics of a mixture. There is evidence that the optimum amount of −200 material is between 3% and 6% for well-graded bituminous concrete.

3. Mixtures containing crushed aggregate have better resistance to rutting than those containing little or no crushed aggregate.

4. Further study, including some very basic research, is needed in order to better understand the mechanics of what happens with various types and amounts of mineral filler in a bituminous mixture.

Acknowledgment

Grateful appreciation is expressed to the Waterways Experiment Station for assistance in the preparation of the manuscript. The Office, Chief of Engineers, is acknowledged for funding much of the work that produced this paper and for granting permission to publish the paper. The authors are also indebted to the Mississippi State Highway Department and the Louisiana Department of Transportation as indicated in the references.

References

[*1*] McRae, J. L, "Theoretical Aspects of Asphalt Concrete Mix Design," *Proceedings,* 3rd Paving Conference, Civil Engineering Department, The University of New Mexico, Albuquerque, NM, 1965.

[*2*] Neepe, S. L., "Mechanical Stability of Bituminous Mixtures: A Summary of Literature," *Proceedings*, American Association of Paving Technologists, Vol. 22, 1953.

[*3*] Bencowitz, I., "Suggested Procedure in Design of Sheet Asphalt," *Industrial and Engineering Chemistry,* Vol. 29, 1937.

[*4*] Brown, E. R., "Evaluation of Asphaltic Concrete Pavements Subjected to T-38 Aircraft Loadings," Masters thesis, Mississippi State University, State College, MS, Dec. 1974.

[*5*] "The Marshall Method for the Design and Control of Bituminous Paving Mixtures," 3rd rev., Marshall Consulting and Testing Laboratory, Jackson, MS, 1949.

[*6*] Crawley, A. B. and McRae, J. L., "The Use of a Gyratory Testing Machine for Controlling the Quality of Hot Plant-Mixed Bituminous Pavements," report presented to Committee D-4, ASTM, Nashville, TN, Dec. 1975.

[*7*] Monismith, C. L. and Vallerga, B. A., "Relationship Between Density and Stability of Asphalt Paving Mixtures," *Proceedings,* Vol. 25, 1956.

[*8*] Arena, P. J., "Evaluation of The Gyratory Compactor For Use in Designing Asphalt Concrete Mixtures," Research Report No. 26, Louisiana Department of Highways, Research and Development, in cooperation with Department of Transportation, Federal Highway Administration, Dec. 1966.

[*9*] Brown, E. R., "Potential for Using Loess in Sand Asphalt Mixture," Miscellaneous Paper GL-85-23, U.S. Army Engineer Waterways Experiment Station, Sept. 1985.

DISCUSSION

Richard L. Davis (*written discussion*)—In the last five years I have seen increasing signs of distress in asphaltic concrete pavements under heavy truck traffic. After study of this problem, I think that a considerable percentage of these failures are due to increased stresses: increased tire pressures causes the pavements to shear. The failure in shear increases air voids and pore size, allowing water to enter the pavements. These shear failures are often labeled as stripping or water problems when they are due to insufficient strength in shear.

"Effect of Aggregates on Performance of Bituminous Concrete," by E. R. Brown, John L. McRae, and Alfred B. Crawley, quite correctly points out that increasing the maximum

[1] Koppers Company, Inc., 1177 Pinewood Drive, Pittsburgh, PA 15243.

size of the aggregate produces a higher stability and requires less asphalt in a given mix. The paper goes on to say that the compacted lift thickness should be at least two times the maximum aggregate size. While this statement is in accordance with conventional wisdom, I believe that we must reexamine such restrictions if we are to cope with our shear problems.

In the ASTM Symposium on Rheology in December 1985, I gave a paper in which I called for larger maximum size stone and an effort to increase the maximum size of the aggregate to more than two thirds the compacted lift thickness. Thousands of miles of pavements with the top size aggregate greater than two thirds of the compacted lift thickness were placed in the early part of this century. These pavements had such high stabilities that they could be compacted to about 2% air voids during construction. When some of these pavements were studied after more than 50 years of service, their air voids were still about 2%. There was no evidence of bleeding in these pavements.

I believe that we could gain in durability and stability of pavements if we could learn to place pavement lifts which have maximum size aggregate which approaches the thickness of the lift.

E. R. Brown, John L. McRae, and Alfred B. Crawley (*authors' closure*)—The authors believe that increasing the maximum aggregate size of well graded mixtures to more than two thirds the compacted lift thickness would be detrimental to the placement and compaction of the asphalt concrete. Pavers that are used today simply cannot place a well-graded mixture to that thickness without tearing and pulling the mat. Compaction of these relatively thin lifts is also difficult due to bridging of the aggregate.

Ervin L. Dukatz, Jr.[1]

Use of Thin Asphalt Surfaces Over Aggregate Base Course for Heavy-Axle Truck Loads

REFERENCE: Dukatz, E. L., Jr., **"Use of Thin Asphalt Surfaces Over Aggregate Base Course for Heavy-Axle Truck Loads,"** *Implication of Aggregates in the Design, Construction, and Performance of Flexible Pavements, ASTM STP 1016,* H. G. Schreuders and C. R. Marek, Eds., American Society for Testing and Materials, Philadelphia, 1989, pp. 64–77.

ABSTRACT: Crushed-stone base on roadways provides structural integrity, facilitates drainage, and minimizes construction costs. At this time, however, it is not being used to its full potential in pavement construction: recent laboratory tests show that a properly designed and constructed crushed-stone base is much stronger than widely believed.

A test pavement project in North Carolina involving the Vulcan Materials Company (VMC), the North Carolina Aggregates Association (NCAA), and the North Carolina Department of Transportation has been undertaken to provide data on the effectiveness of pavements constructed with thin asphalt surfaces over thick aggregate base courses. State Route 1508 in North Carolina, constructed in early 1985 and designed to have a five-year life, is one of the test roads used in the project. The road has been subjected to 60% of its design traffic in 18 months and shows minimal amount of distress. Because it is trafficked by all incoming and outgoing aggregate haulers from a VMC quarry (for which it was constructed), SR 1508 is of interest to engineers concerned about the deformation of pavements (rutting) due to the loadings of heavy truck traffic.

KEY WORDS: bituminous concrete, flexible pavements, aggregate base course, permeability

Crushed-stone base overlain with asphalt or portland cement concrete serves as an integral part of a roadway in providing structural integrity and, in many cases, drainage, and minimizes construction costs. At the present time, however, crushed stone is frequently not being used to its fullest advantage in pavement construction, nor is it always given, the full structural credit to which it is entitled [*1,2*]. Recent laboratory and field tests have shown that a properly designed and constructed crushed-stone base is sometimes stronger than American Association of State Highway and Transportation Officials (AASHTO) Test Road results would indicate. The North Carolina Aggregates Association (NCAA), along with the North Carolina Department of Transportation (NCDOT), has undertaken a test pavement project to investigate the validity of design coefficients developed at the AASHTO Test Road when utilized with construction materials typical to the North Carolina area. The original proposal included five test sections. These test sections were chosen so that the performance of equivalent design thicknesses of crushed-stone base material with asphaltic concrete base materials could be compared in terms of support values, resistance to de-

[1] Senior materials engineer, Vulcan Materials Co., Birmingham, AL 35253-0497.

flection, and distortion offered by the thoroughly compacted crushed stone and asphaltic base materials. However, due to a number of considerations, these test sections were reduced to one half-mile (0.8 km) test section on State Route (SR) 1508 located in Vance County. This section of SR 1508 serves as a local access and haul road for Greystone quarry of Vulcan Materials Company (VMC). The upper portion of the road also services an asphalt plant. The section of SR 1508 chosen was composed of various stone layers and asphalt surfacings (Fig. 1). NCDOT coring records indicated that the subgrade soil consisted of A-4 to A-7 materials. The roadway was originally surfaced with asphalt in 1946. Maintenance records did not indicate any resurfacing being placed over the original pavement. However, during the reconstruction, evidence was found of three layers of asphalt and crushed stone in the old roadbed. The pavement conditions survey performed in the spring of 1984 by the NCDOT maintenance unit rated the road as needing reconstruction. The original road, which had between 2.5 and 3.8 cm (1 and 1.5 in.) of surface treatment over 30.5 to 35.6 cm (12 to 14 in.) of aggregate base course (ABC), had a design life of from two to five years—this for a road that was placed in 1946. Approximately 38 years later, that road was still supporting traffic, although it was showing major severe cracking in the exit lane from the quarry and some areas of severe block cracking in the entrance lane to the quarry. The preconstruction survey noted that only one section has been patched. The patch was between Stations 11+60 and 11+80 in the exit lane, 2.5 m (5 ft) in width beginning at pavement edge. Due to heavy traffic from the quarry, the pavement has sunk 6.3 to 7.6 cm ($2\frac{1}{2}$ to 3 in.) from the quarterpoint to the edge of pavement in both lanes.

Based on the subgrade survey and traffic records, the NCDOT pavement design section projected that the road should be based on 163 daily 8165-kg (18 kip) equivalent axle loads (EAL's). For the purposes of this project, the NCAA proposed a pavement design of 5.1 cm (2 in.) of an NCDOT I-1 surface asphalt hot-mix [3] and 33 cm (13 in.) Aggregate Base Course (ABC) material for the section between Stations 5+0 and Stations 14+50 (Fig. 2). The projected life of this section is five years. The remaining sections of the road which

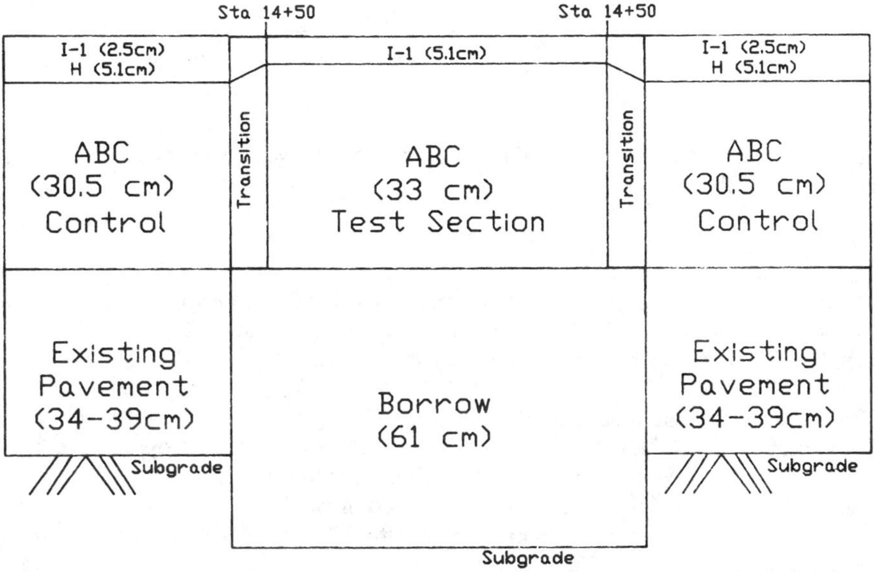

FIG. 1—*Elevation view of existing, control, and test road sections.*

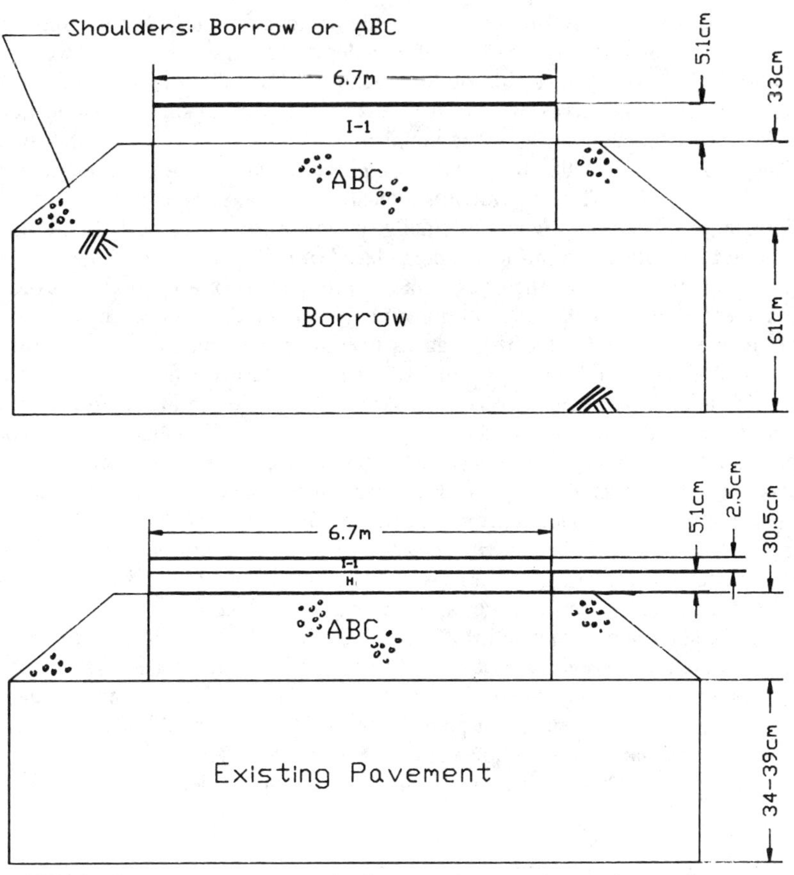

FIG. 2—*Pavement cross section:* (top) *test road section,* (bottom) *control section.*

were rebuilt received 2.5 cm (1 in.) of I-1, 5.1 cm (2 in.) of NCDOT H binder [3], and 30.5 cm (12 in.) of ABC (Fig. 2).

Construction

The project was constructed by North Carolina Division 5 Maintenance Forces according to NCDOT Specifications [3] and normal construction procedures. The project was constructed under traffic with minimal rerouting of the heavy truck traffic coming from the quarry.

The old asphalt, stone base, and 61 cm (24 in.) of subgrade was removed in the section. The upper 30.5 cm (12 in.) of stone was stockpiled for use on other projects. The bottom 30.5 cm (12 in.) of subgrade was cast to the side as shoulder material.

The borrow material for the subgrade was placed in two 30.5-cm (12-in.) lifts and compacted with a sheepsfoot roller. Compaction was to 95% of AASHTO T-180 for the first lift and 100% for the second lift.

During construction a number of water problems were encountered. Between Stations 12 and 13, saturated subgrade materials were found. These materials were removed and

TABLE 1—*Laboratory subgrade properties.*

Soil Type	CBR	Station	Depth	
			cm	(in.)
A-4 (1)	11	13+90	7.6 to 12.7	(3 to 5)
A-7 (5)	19	16+15	7.6 to 12.7	(3 to 5)
A-7 (7)	20	20+50	5.1 to 10.2	(2 to 4)

replaced with the A-7 borrow material. Another problem during construction was rain interrupting construction of the borrow sections for as long as four days. After each rain stoppage, a sheepsfoot roller was used to dry the subgrade.

After the borrow material was compacted, shoulder materials were placed and compacted. Stockpiled ABC material was used to construct shoulders between Stations 10+0 to 12+0. This ABC was not from the same approved stockpile used for the base. The shoulders for the remainder of the project were constructed from the excavated material.

Material Characteristics

Subgrade

The NCDOT made a number of tests to characterize the subgrade soil. Table 1 gives the subgrade soil laboratory test results. These results show that the California Bearing Ratio (CBR) of the subgrade ranges from 11 to 20. *In situ* subgradc properties are given in Table 2. The field CBR values range from a low of 13 at Stations 8+00 and 10+00 to 24 Station 12+00. As the old road was taken up, additional subgrade soil sampling was done. This sampling indicated that the soil ranged from an A-4 at the beginning of the project to an A-7 (14) at the end of the project at Station 14+50. The gradation of the borrow material is shown in Fig. 3.

The base ABC (from NCDOT approved stockpile) was placed in two layers, 15.2 and 17.8 cm (6 and 7 in.), and compacted by a vibratory roller and a rubber-tired roller. All lifts were compacted to 100% AASHTO T-180. During placement of the ABC, traffic was detoured until the paving was completed.

The Type H binder mix was placed outside the test section in the morning. That afternoon the total project was paved with Type I-1. Design compaction was to 98% of maximum test

TABLE 2—In situ *subgrade properties.*[a]

Station No.	Field Moisture, %	Optimum Moisture, %	In-Place Dry Density, kg/m³	(lb/ft³)	Maximum Dry Density, %	Percent Compaction	Field CBR at 0.1	Field CBR at 0.2
6+00	17.2	22.0	1590	(99.4)	101.1	98.3	22	20
8+00	...	...	...	...	...	96.0	13	17
10+00	15.8	22.0	1624	(101.5)	101.4	100.1	13	13
12+00	...	...	...	...	...	104.9	24	24
14+00	18.6	21.5	1640	(102.5)	102.3	100.2	13(14)	17(18)

[a] The Field Moistures, Optimum Moistures, In-Place Dry Densities, Maximum Dry Densities, and Percent Compaction at Stations 6+00, 10+00, and 14+00 were determined in accordance with NCDOT Density Test No. 1. The Percent Compaction at Stations 8+00 and 10+00 was determined in accordance with NCDOT Density Test No. 1-A. Both procedures use estimated optimum moisture.

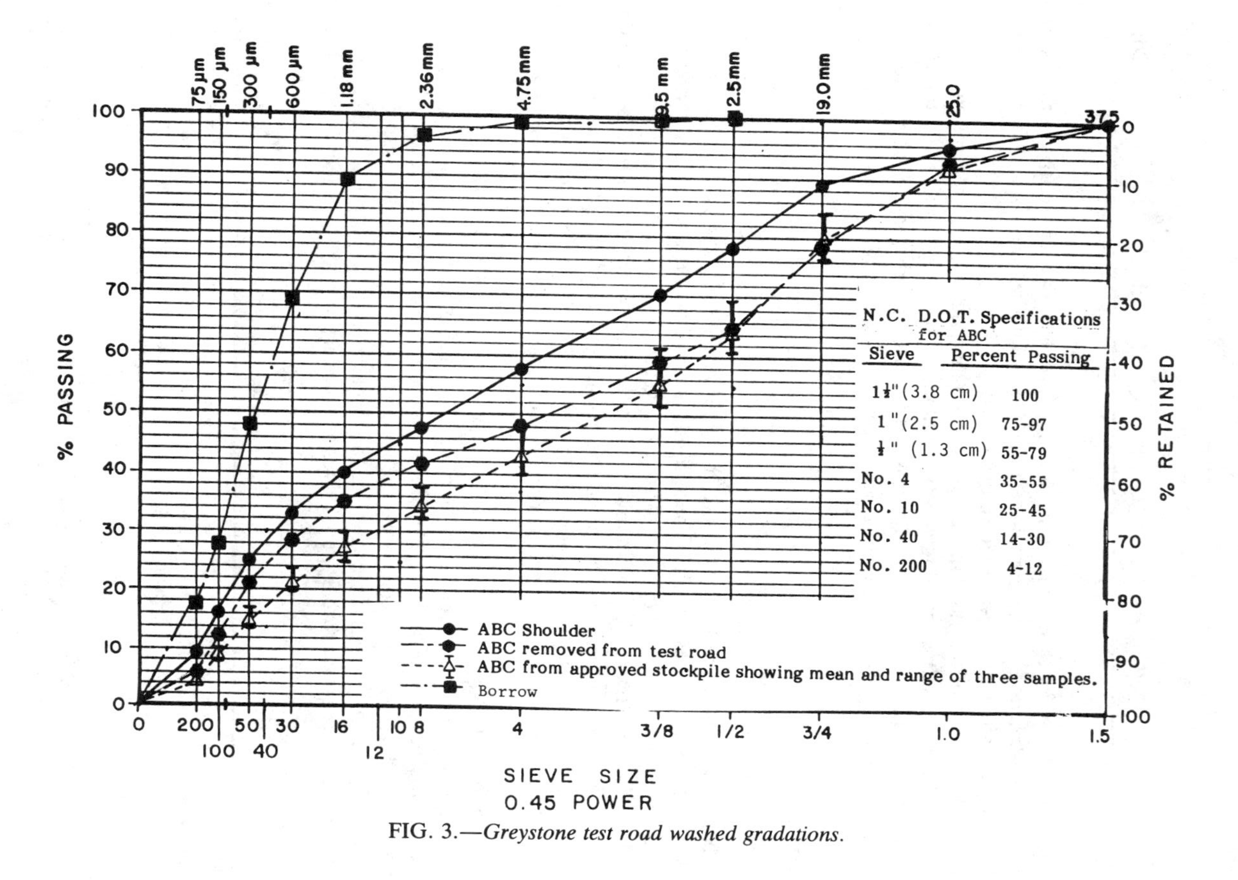

FIG. 3.—*Greystone test road washed gradations.*

TABLE 3—*Greystone test road—base and subgrade material test results.*

Test Property	Borrow	NCDOT Approved ABC Stockpile	ABC Removed	ABC Shoulder
I. ATTERBERG LIMITS				
LL	38.0	. . .	. . .	. . .
PI	NP	NP	NP	NP
II. OPTIMUM DENSITY (AASHTO T-180)				
Dry density				
kg/m³ (lb/ft³)	1920 (120.0)	2336 (146.0)[c]	2368 (148.0)	. . .
Optimum moisture, %	15.5	5.8	6.0	. . .
III. PERMEABILITY				
m/day (ft/day)				
K, at optimum density	NM[b]	NM	NM	NM
K, at (density)[c]	0.003 (0.01)	0.08 (0.28)	0.03 (0.10)	0.04 (0.14)
kg/m³ (lb/ft³)	1376 (86)	1856 (116)	1856 (116)	1888 (118)
IV. AASHTO SOILS CLASSIFICATION				
Group	A-2-4(0)	A-1a(0)	A-1a(0)	A-la(0)
Description	yellowish-brown silty sand	stone	stone	stone

[a]NCDOT design optimum density = 2249.6 kg/m³ (140.6 lb/ft³).
[b]NM = nonmeasurable.
[c]Based on material passing 1.3-cm (½-in.) sieve.

strip density. Actual compaction values averaged 96% of test strip density and the pavement thickness averaged 4.4 cm (1¾ in.).

Table 3 is a summary of test results obtained at the VMC Research and Development Laboratory in Birmingham. Subgrade specimens taken on site show the plasticity index to range from 0 to 6%. The optimum densities determined by AASHTO T-180 compaction at the VMC Research and Development Laboratory in Birmingham and those determined by the NCDOT indicate that the dry density of the borrow and subgrade soil is 1920 kg/m³ (120 lb/ft³) at an optimum moisture content of approximately 16%.

Aggregate Base Course

The ABC properties results are presented in Table 3. The optimum dry density for ABC from the NCDOT-approved stockpile was 2249.6 kg/m³ (140 lb/ft³) (determined by NCDOT) at a moisture content of 5.8%. The existing ABC had an optimum dry density of 2368 kg/m³ (148 lb/ft³) at a moisture content of 6.0%. An in-place dry density measured by the NCDOT after two years of traffic was 2304 kg/m³ (144 lb/ft³), with an *in situ* moisture content of 2.4%. These results indicate that even after compaction to AASHTO T-180, the base material experienced further compaction due to traffic. The gradation of the ABCs are shown in Fig. 3. The ABCs used for the shoulders and for the base course are well-graded, containing a maximum aggregate size of 3.8 cm (1½ in.). The shoulder ABC contains approximately 10% passing the No. 200, and the base ABC contains 4% passing. The implications of the differences in the amount of No. 200 are discussed under permeability.

Pavement Surface

An NCDOT I-1 plant mix was used as a surface course. This mix was 100% passing the 1.3-cm (1½-in.) sieve with an asphalt cement content of 6.1%. The complete design is given in Table 4.

TABLE 4—*I-1 asphalt surface design for SR 1508.*

	Sieve	Percent Passing: Typical Gradation, %	Percent Passing: JMF
1. Gradation			
	1.9 cm (¾ in.)	100	100
	1.3 cm (½ in.)	100	97 to 100
	No. 4	70	61 to 75
	No. 8	53	45 to 55
	No. 40	26	19 to 29
	No. 80	13	8 to 18
	No. 200	5	3 to 7
2. Marshall properties (50-blow design)			
Optimum % AC	6.1		
Marshall stability, kg(lb)	843.7 (1860)		
Marshall flow, 0.025 cm (0.01 in.)	23 (9)		
Air voids, %	4.6		
Voids filled, %	75		
Unit weight, kg/m^3 (lb/ft^3)	2304 (144)		

Permeability

On one side of the test road section was a drainage ditch (removed during reconstruction) receiving discharge from the quarry, and on the other side of the test road is a large pond. Many of the people involved with the project were concerned about water being trapped in the base. This concern was based on the anticipated lower permeabilities of the shoulder and subgrade materials, which contain more No. 200 material than the base. Permeability tests were conducted to investigate this concern. The permeabilities were nonmeasurable in the borrow material and in the base materials in tests conducted according to the AASHTO Standard Method of Test for Permeability of Granular Soils (Constant Head) (T-215-70), using a 15.2-cm-diameter (6-in.) constant-head permeameter at optimum density. The material was re-sieved, and those materials passing through a 1.3-cm (½-in.) sieve were retested. The specimens were compacted to approximately 80% of optimum and tested in a 2.5-cm-diameter (1-in.) falling-head permeameter constructed at the VMC Laboratory. The results presented in Table 3 show that the subgrade material compacted to a density of 1376 kg/m^3 (86 lb/ft^3) was the least permeable material, with a coefficient of permeability of 0.003 m (one hundredth of a foot) per day, and that the stockpiled ABC material was the most permeable at 0.08 m (0.28 ft) per day. The ABC shoulder material had a permeability of 0.04 m (0.14 ft) per day, which is half of that of the base ABC. All permeability values were much less than 304.8 m (1000 ft) per day, the value recommended by the National Stone Association for draining base courses. As one would expect, the ABC from the approved stockpile had the coarsest gradation and the borrow material had the finest gradation. These results would indicate that water infiltrating into the base would become trapped there by the less permeable subgrade and shoulder materials.

Pavement Performance

Rutting History

Summarized in Figs. 4 to 7 are the results of rut measurements taken at the test road. These figures show that rut depths held very constant over the life of the project except for the outbound outer wheel path around Station 10+00. This is the exit lane from the quarry.

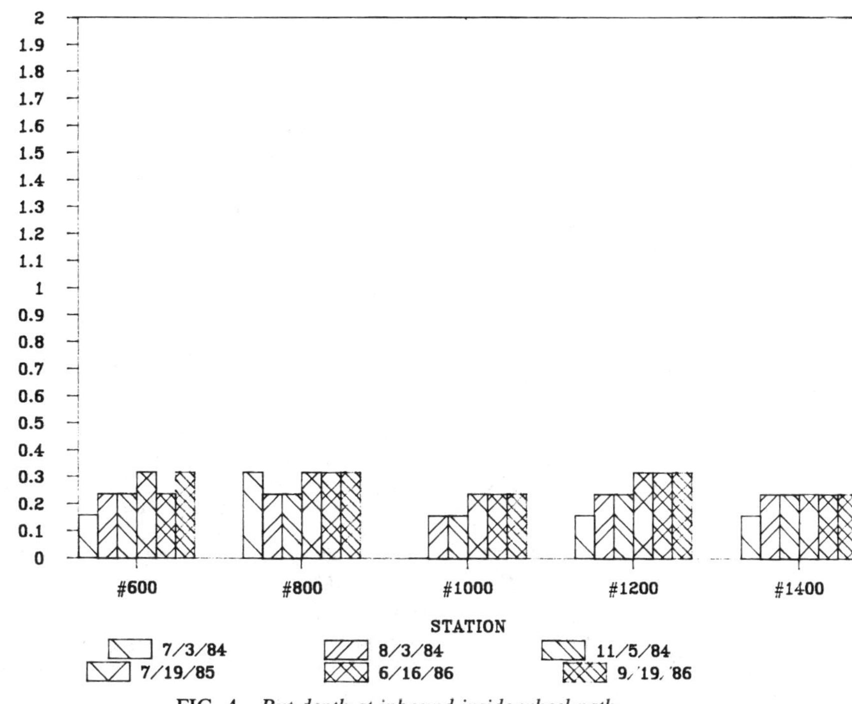

FIG. 4—*Rut depth at inbound inside wheel path.*

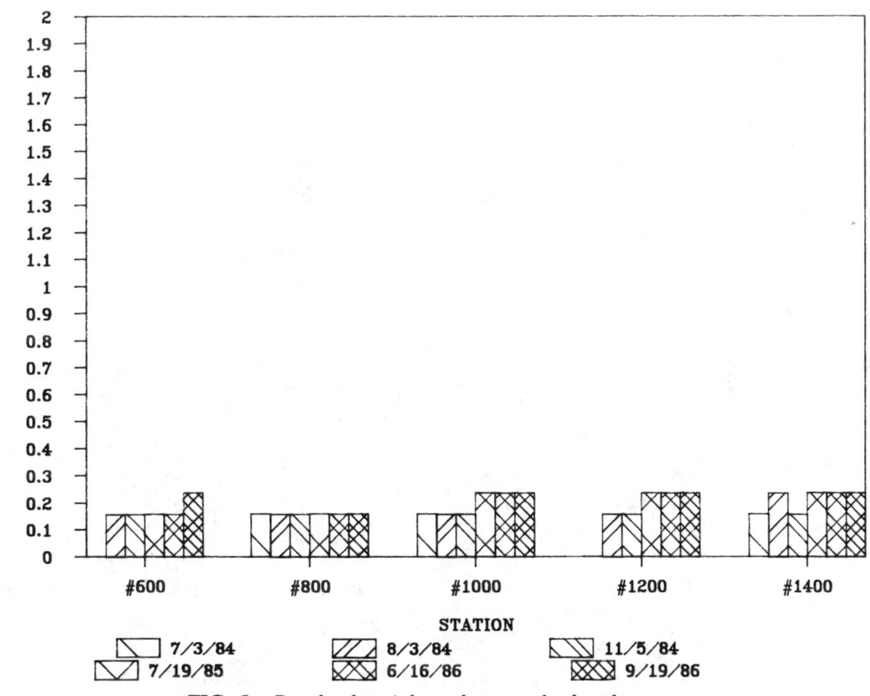

FIG. 5—*Rut depth at inbound outer wheel path.*

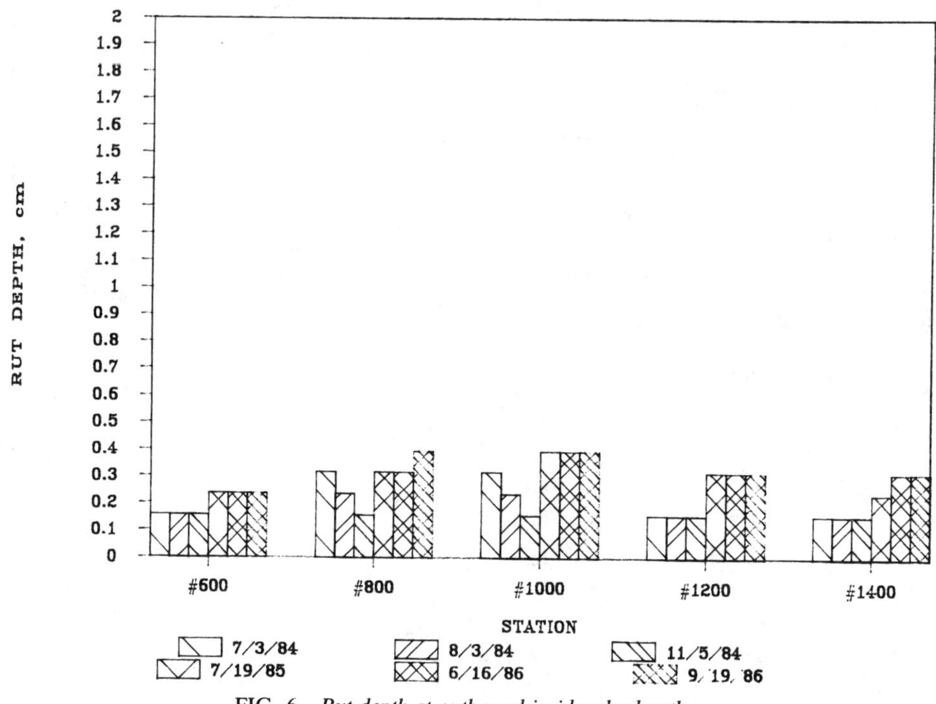

FIG. 6—*Rut depth at outbound inside wheel path.*

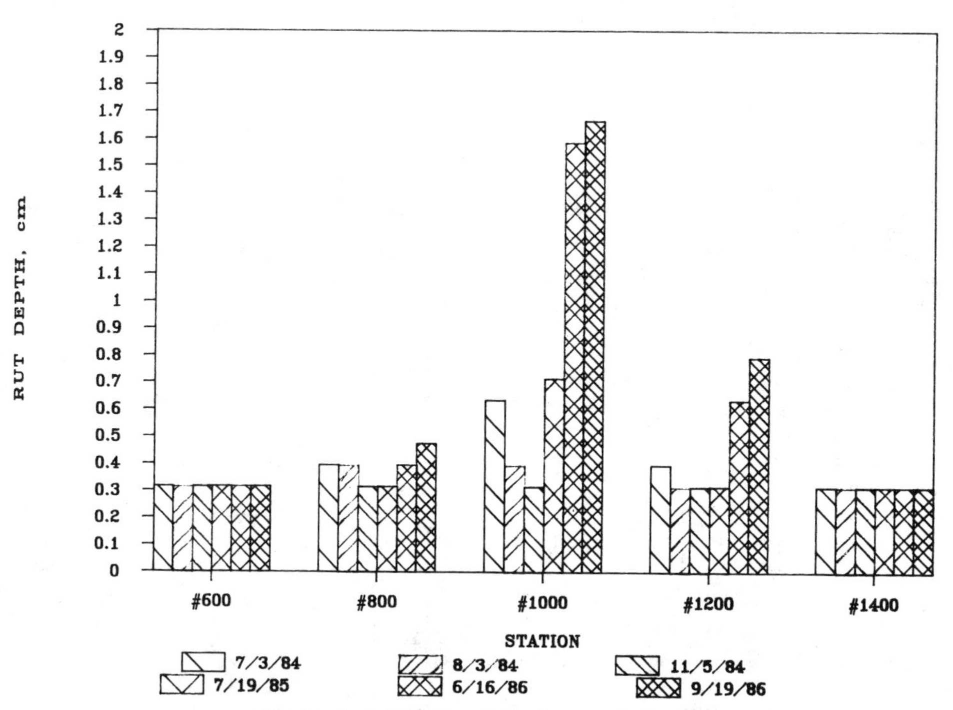

FIG. 7—*Rut depth at outbound outer wheel path.*

The outer wheel path between Stations 10+00 and 12+00 is showing extensive alligator cracking after two years of traffic. Figure 7 illustrates the increase in rut depth between the summers of 1985 and 1986. Between these two stations are the areas in which there were problems with moisture during construction (saturated subgrade) and also the area that had previously been patched in the original road. Also, during the summer of 1985, the area around Station 10+00 had shown signs of large volumes of water flowing over and under the pavement section. This was evidenced by washout areas on either side of the pavement.

Because the rest of the pavement is doing well under much higher than anticipated axle loadings, a decision was made to put a nonstructural patch on the alligator cracked area. The patch is to prevent further moisture infiltration and to smooth the pavement surface.

Before the patch was applied, the affected area was trenched in an attempt to determine the cause of the alligator cracking. The results of in-place testing are presented in Table 5.

The subgrade which was placed at close to the 15.5% optimum moisture content now has between 22 and 24% moisture. At station 10+00, where the worst cracking occurs, the CBR has decreased from between 11 and 13% to between 7 and 8%. Visual observation of the pavement cross section exposed by the trench shows that there is no contamination of the base by the subgrade. There is a clear line of demarcation between each pavement layer. Thus, the cracking and rutting appear to be due to reduced subgrade support. The I-1 mix averaged less than 3.8 cm (1½ in.) in the damaged section. So the asphalt surface was not contributing its design support value. However, because the half-lane trench posed more questions than answers, plans are being made to trench across both lanes.

An additional factor that needs to be taken into account is that the road had been striped so that the exit lane was a foot narrower than the entrance lane. This resulted in higher than anticipated loads on the outside edge of the pavement. The shoulder and the edge of the pavement showed evidence of one tire of the dual-axle being on the pavement and the other tire either riding above or going through the shoulder. This is believed to be part of the cause for some of the longitudinal cracking seen in the outside wheel path on the exit lane.

Equivalent Axle Loads

Traffic records were kept by the VMC weigh clerk at the Greystone quarry. These records included the number, type, and gross loads of the vehicles utilizing the test pavement area.

TABLE 5—*Trenching specimen results.*

Station	ABC 10+20	Subgrade 11+50	11+50	10+10	10+10
Distance from edge of pavement, m (ft)	0.91 (3)	0.91 (3)	1.98 (6.5)	1.07 (3.5)	1.98 (6.5)
Lane	WBL[a]	WBL	WBL	WBL	WBL
CBR Values at					
0.04 cm (0.1 in.)	...	10.4	7.1[b]	6.6	7.1
0.08 cm (0.2 in.)	...	11.6	9.5[b]	7.3	8.5
0.12 cm (0.3 in.)	...	11.8	10.7[b]	7.5	8.5
Percent moisture	2.4	21.0	24.3	22.4	22.2
Wet density, kg/m³ (lb/ft³)	2352 (147.0)				
Dry density, kg/m³ (lb/ft³)	2297.6 (143.6)				
AASHTO T-180 density, (laboratory unit weight)	2249.6 (140.6)				
Percent compaction	102.1				

[a]Westbound lane.
[b]Data listed are corrected for surface disturbances.

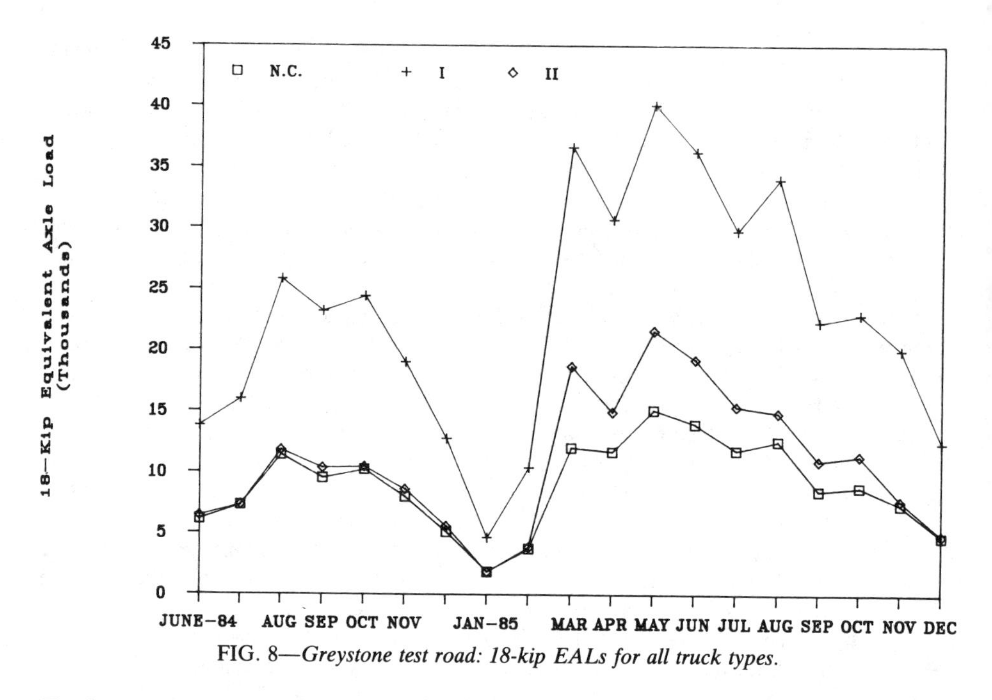

FIG. 8—*Greystone test road: 18-kip EALs for all truck types.*

Truck measurements were used to calculate the accumulative EALs for the pavement section. Comparison of the calculated EALs from the actual truck traffic with the design values gives an indication of performance. Little or no distress while supporting loads in excess of the design 163 daily EALs would be an indication of good performance. The cumulative EALs were calculated in three ways: (1) by an algorithm developed by the NCDOT, (2) by AASHTO procedures, and (3) by a modified AASHTO procedure [*4*].[2] The modified AASHTO procedure assumed that 4082 kg (9 kips) of the gross load was carried on the steering axle with the remainder on the other axles. As shown in Fig. 8, this method compared very well with the North Carolina procedure. The AASHTO procedure of distributing the gross weights over the rear axles gave EALs that were two to three times higher. Each method used showed that the pavement had received more than 163 daily EALs and was still performing well except in the noted section. A summary of the EALs determined by each method is given in Table 6. Averaging the NCDOT results and the modified AASHTO results gave an average daily EAL of 342, which is 210% of design. This is outstanding performance.

Discussion of Results

Based on these results, it does not appear that the low permeability of the borrow material used for shoulders has been a factor affecting the performance of this pavement. The largest factors appearing to affect the performance of the pavement is the presence of water between Stations 10+0 and 12+0. From the EAL calculations, the pavement is performing very well, receiving over 60% of its design traffic in 18 months. Using the design traffic figure, one can determine the structural coefficient for the base course. If we assume that the asphaltic surface coefficient is 0.44, the coefficient for the base course is backcalculated from

[2] Ackerman, M., North Carolina Department of Transportation, personal communication, October 1986.

TABLE 6—*Summary EAL loads for SR 1508.*

North Carolina DOT Method

Month	Pickup	Single Axle	Tandem	Triple Axle	Tractor Trailer	Total
June 1984	0.56	622.22	2216.36	2502.58	771.76	6113.48
July	0.31	768.71	2348.69	3188.38	974.38	7280.46
Aug	0.46	1469.09	3141.19	4721.10	2062.96	11394.79
Sep	0.15	1539.38	2893.91	2919.11	2153.39	9505.94
Oct	0.40	1914.76	3218.20	3418.56	1676.99	10228.90
Nov	0.27	1120.93	2692.46	2669.74	1467.04	7950.43
Dec	0.23	1061.48	1528.48	1262.74	1232.47	5085.39
Jan 1985	0.32	486.17	630.61	485.90	248.99	1851.99
Feb	0.42	1294.96	1501.65	551.55	388.60	3737.18
Mar	0.45	1747.63	2324.89	3395.81	4554.26	12023.03
Apr	0.20	1372.61	3042.44	2527.19	4754.34	11696.78
May	0.19	682.46	2631.50	3132.34	8650.83	15097.32
Jun	0.51	913.65	2567.49	3842.15	6606.96	13930.77
Jul	0.27	865.08	2640.06	3550.67	4738.04	11794.14
Aug	0.45	3148.48	2527.00	2543.96	4321.89	12541.78
Sep	0.20	1114.25	1754.91	2012.83	3605.43	8487.62
Oct	0.33	1064.21	1916.12	2175.28	3642.89	8798.83
Nov	0.44	2552.78	1676.87	1534.10	1574.58	7338.76
Dec	0.35	1512.14	1061.89	1308.09	848.01	4730.47
Total	6.50	25250.98	42314.71	47742.08	54273.80	169588.07
						(57%, 2.85 years)

I. AASHTO Method

Month	Pickup	Single Axle	Tandem	Triple Axle	Tractor Trailer	Total
June 1984	2.81	2003.40	5904.97	3613.74	2311.17	13836.09
July	1.66	2472.97	6105.66	4482.62	2930.80	15993.70
Aug	2.41	4726.21	8192.27	6658.49	6249.25	25828.62
Sept	0.83	4949.69	7570.56	4168.64	6499.69	23189.42
Oct	2.22	6154.59	8426.40	4829.02	5028.41	24440.64
Nov	1.49	3603.11	7068.26	3857.34	4456.13	18986.33
Dec	0.83	3414.44	3987.68	1772.42	3585.61	12760.98
Jan 1985	1.60	1559.57	1647.59	699.97	746.91	4655.63
Feb	2.26	4191.53	3924.14	792.28	1528.51	10438.72
Mar	2.37	5872.80	6931.86	5580.49	18283.88	36671.40
Apr	1.04	4652.97	8008.24	3588.67	14435.81	30686.75
May	1.07	2301.52	6886.84	4444.55	26477.23	40111.20
June	2.88	3054.03	6756.99	5671.15	20750.36	36235.41
July	1.48	2847.73	6893.82	5365.35	14655.44	29763.82
Aug	2.40	10412.46	6682.78	3634.76	13298.96	34031.36
Sep	1.13	3728.46	4572.70	2899.66	11127.81	22329.75
Oct	1.74	3479.78	5045.63	3189.10	11234.22	22950.47
Nov	2.30	8580.56	4431.15	2211.97	4830.03	20056.00
Dec	1.88	5052.06	2804.09	1922.86	2634.49	12415.39
Total	34.40	83057.87	194933.90	69383.06	171064.72	435381.67
						(146%, 7.3 years)

TABLE 6—*Continued.*

II AASHTO Method

Month	Pickup	Single Axle	Tandem	Triple Axle	Tractor Trailer	Total
June 1984	2.81	355.78	2686.19	2024.60	1342.73	6412.12
July	1.66	450.69	2691.31	2460.78	1711.66	7316.09
Aug	2.41	843.59	3619.43	3670.43	3670.54	11806.40
Sept	0.83	889.07	3354.83	2318.23	3801.17	10364.12
Oct	2.22	1116.63	3737.32	2668.80	2932.94	10457.90
Nov	1.49	662.34	3141.45	2162.78	2626.90	8594.96
Dec	0.83	621.87	1762.76	972.14	2205.01	5562.61
Jan 1985	1.60	285.99	730.83	391.33	436.22	1845.98
Feb	2.26	768.64	1733.96	442.48	917.70	3865.05
Mar	2.37	1090.62	3265.29	3194.24	11115.53	18668.05
Apr	1.04	878.27	3560.64	1992.78	8503.11	14935.85
May	1.07	437.76	3060.29	2468.05	15646.82	21614.00
June	2.88	566.84	3053.25	3205.69	12405.02	19233.67
July	1.48	526.90	3052.22	3061.60	8727.19	15369.40
Aug	2.40	1931.46	3030.50	2022.35	7880.60	14867.31
Sept	1.13	692.40	2024.56	1619.44	6610.54	10948.06
Oct	1.74	643.85	2267.24	1797.43	6657.09	11367.35
Nov	2.30	1606.76	2012.92	1240.36	2856.52	7718.86
Dec	1.88	937.68	1269.35	1084.74	1572.09	4865.73
Total	34.40	15307.15	50054.36	38798.25	101619.37	205813.527 (69%, 3.45 years)

[a]Method II: For front axle, use 9 kips divided by the number of loads to find equivalency factor. For the rear axle, use the gross weight minus 9 kips, divided by number of loads. The EAL is the sum of front and rear values.

a design SN = 2.70 to be between 0.18 and 0.20. This assumes the I-1 surface is a full 5.08 cm (2 in.). This is a substantial increase over the 0.14 which is currently being used as the structural coefficient number. Construction reports and data from the test trench indicate that the I-1 surface is less than 5.08 cm (2 in.). That makes the backcalculated base coefficients conservative estimates.

Conclusions and Future Work

From this preliminary study on SR 1508 in Vance County, North Carolina, it can be concluded that a pavement consisting of a thin layer of asphalt surface over a thick layer of aggregate base course works very well. Further, the structural coefficient for the aggregate base course appears to be 0.18 or higher, instead of the 0.14 used for design.

This report is the first of two reports on this particular test section. The second report will contain data from Benkelman beam and falling-weight deflectometer (FWD) tests. It is the intention of the North Carolina Aggregates Association and the North Carolina Department of Transportation to continue monitoring this road until its design life is reached or until it is replaced, whichever comes first.

Acknowledgments

The author wishes to thank the North Carolina Department of Transportation and the North Carolina Aggregates Association for their help.

References

[1] "Fourth Cycle of Pavement Research at the Pennsylvania Transportation Research Facility," Research Project 82-11, FHWA/PA 84-028, Federal Highway Administration, Dec. 1984.

[2] Greene, R. L., "Analysis of An Experimental Pavement at the Vulcan Materials Company Quarry at Stockbridge, Georgia," M.S. thesis, Georgia Institute of Technology, Atlanta, GA, Jan. 1986.

[3] *Standard Specifications for Roads and Structures,* North Carolina Department of Transportation, 1 Jan. 1984.

[4] AASHTO Guide for Design of Pavement Structures, American Association of State and Highway Transportation Officials, Washington, DC, 1986.

Ervin L. Dukatz, Jr.,[1] *and Richard S. Phillips*[1]

Hot-Mix Asphalt Moisture Susceptibility Problems: The Need to Test and Specify via a Common Procedure

REFERENCE: Dukatz, E. L., Jr., and Phillips, R. S., **"Hot-Mix Asphalt Moisture Susceptibility Problems: The Need to Test and Specify via a Common Procedure,"** *Implication of Aggregates in the Design, Construction, and Performance of Flexible Pavements, ASTM STP 1016,* H. G. Schreuders and C. R. Marek, Eds., American Society for Testing and Materials, Philadelphia, 1989, pp. 78–95.

ABSTRACT: One of many problems facing the hot-mix asphalt designer, specifier, and producer is how to properly predict the behavior of the asphalt concrete when exposed to moisture. At present, a broad spectrum of accelerated test procedures are available to the engineer that to some degree "predict" the moisture susceptibility of the mix. These procedures include the Lottman, the modified Lottman, and the boiling water stripping test. With the number of procedures available, and with different procedures being specified in different parts of the country, the engineer must answer the question of which procedure best suits the prediction of future performance.

Once a procedure is chosen, then the method by which the data are interpreted becomes important. For the Lottman procedures, splitting tension specimens are often prepared at 7 ± 1% air voids. Grossly different and misleading results will be obtained if conditioned specimens having 8% air voids (7 + 1%) are tested and compared to control specimens having 6% air voids (7 − 1%). While the preceding combination would suggest less than actual performance, the opposite is possible if the extremes are reversed. Either case is undesirable and can prove costly if a local aggregate is excluded for a false low value or if a poor material is accepted for a false passing value.

The authors suggest that a better procedure is to prepare specimens representing the low (4 to 6%), midpoint (6 to 8%), and high (8 to 10%) end of the air void range allowed for the test. This should be done for both the conditioned and the control specimens. After determination of the splitting tensile strength for each specimen, the results should be plotted as tensile strength (S_t) versus percent air voids for each condition. The strength at 7% voids for each test condition can be determined by using graphical interpolation and a valid comparison can be made for all conditions.

Another suggestion of the authors is that a minimum conditioned tensile strength at 7% air voids be specified in conjunction with a minimum retained tensile strength ratio (TSR). The TSR requirement may be waived for treated asphalt mixes in which the conditioned tensile strength exceeds the minimum specified tensile strength and is also higher than the tensile strength of the unconditioned untreated mix.

KEY WORDS: moisture susceptibility, Lottman, modified Lottman, tensile strength, tensile strength ratio (TSR), air voids, Marshall Method, antistrip, stripping

A frequently observed problem by state highway agencies is moisture damage. Moisture damage is defined as any problem that is induced by moisture, including stripping. Research

[1] Senior materials engineer, Construction Materials Group, and laboratory supervisor, CMG Research and Development Laboratory, respectively, Vulcan Materials Company, P.O. Box 7497, Birmingham, AL 35253.

[1,2] has found that moisture damage is related to many factors, including aggregate type, asphalt cement source and grade, mix design construction, climate, and test procedure. To identify mixes prone to moisture damage and thus reducing its effects, many states are using some type of moisture susceptibility test. Two of the more prevalent test methods are the Lottman procedure [3] and the modified Lottman, also known as the Tunnicliff and Root procedure [4]. Both of these methods involve conditioning test specimens by vacuum saturation and a 60°C (140°F) soak. The two procedures differ in that the modified Lottman does not include the Lottman freeze cycle and uses both a higher test temperature and a faster loading rate than does the Lottman. Both procedures determine the number of blows in a Marshall compactor to reach the desired air void content of 7 ± 1% by trial and error.

The trial-and-error period consumes both time and samples until the desired air void parameter is reached. For routine work with the same aggregate, gradation, and asphalt, a minimum of samples can be used for each test. When any component of the asphalt mix is changed, the trial-and-error period must be repeated.

This paper discusses an alternate method of conducting either procedure, one which reduces the number of test samples and improves accounting for random variation and outliers. For this paper, "procedure" refers to sample preparation, that is, either the Lottman or modified Lottman procedures. "Methods" refers to how the data from each procedure are analyzed. The method of data averaging and the alternate method of curved fitting will be compared. Results from the Lottman and modified Lottman procedures will also be compared. Based on these results, a recommendation for a two-tiered acceptance specification will be given.

Scope

The data contained herein were compiled from routine testing at the Vulcan Materials Company Research and Development Laboratory located in Birmingham, Alabama. The conclusions presented in this paper are based on three aggregate types, six aggregate sources, eight aggregate blends, three asphalt sources, eleven antistrip combinations, and two moisture-susceptibility testing procedures. These are presented in Table 1 along with the codes used in the paper.

Background

Vulcan Materials Company operates in 12 states, primarily east of the Mississippi River. Vulcan's Research and Development Laboratory routinely conducts Marshall mix designs and other specification tests such as moisture susceptibility for the areas in which they operate. The Research and Development Lab is charged with the responsibility to provide technical assistance to the more than 100 quarries Vulcan operates. While the test procedures followed were routine, the materials used in each asphalt mix varied. Following either the Lottman or modified Lottman procedure, a large number of Marshall pills had to be prepared for each new asphalt-aggregate combination to determine the number of blows needed during Marshall compaction to reach 7 ± 1% air voids. So that no effort would be wasted, all pills were tested and the resulting data analyzed. Figure 1 displays the results obtained using the Lottman procedure for all materials, while Fig. 2 presents only the data from the modified Lottman. The pills with the highest air voids had the lowest tensile strengths and vice versa. A more detailed analysis was conducted to determine the effect of air voids on the tensile strength ratio (TSR). The TSR is the ratio of average tensile strength after conditioning to that before conditioning expressed in percent. Georgia and Kentucky require surface asphalt mixes to have a TSR ≥80%. For base and binder asphalt mixes, Kentucky requires a minimum TSR of 70% and Georgia a minimum TSR of 80%.

TABLE 1—*Materials used in study—key to mix codes.*

Code	Material
	AGGREGATES (CA)
G	river gravel
L	limestone
S	slag
I	granite
	ASPHALT MIXES (TYPE)
Base	base, coarse
B	binder
E	coarse surface
F	fine surface
	ADDITIVE (AS)
B	0.5% liquid antistrip *x*
D	1.0% hydrated lime + 0.5% liquid antistrip *y*
H	1.0% hydrated lime + 0.5% liquid antistrip *x*
I	1.0% hydrated lime + 0.5% isoprene
K	0.5% liquid antistrip *y*
L	1.0% hydrated lime
N	no additive
Q	0.5% liquid antistrip *z*
S	1.0% liquid antistrip *y*
P	1.0% hydrated lime + 0.5% polypropylene
T	1.0% liquid antistrip *x*
	PROCEDURE (PRO)
Ga	NCHRP 192 (4, 11), as used in Georgia
Ky	NCHRP 274 (5, 12), as used in Kentucky

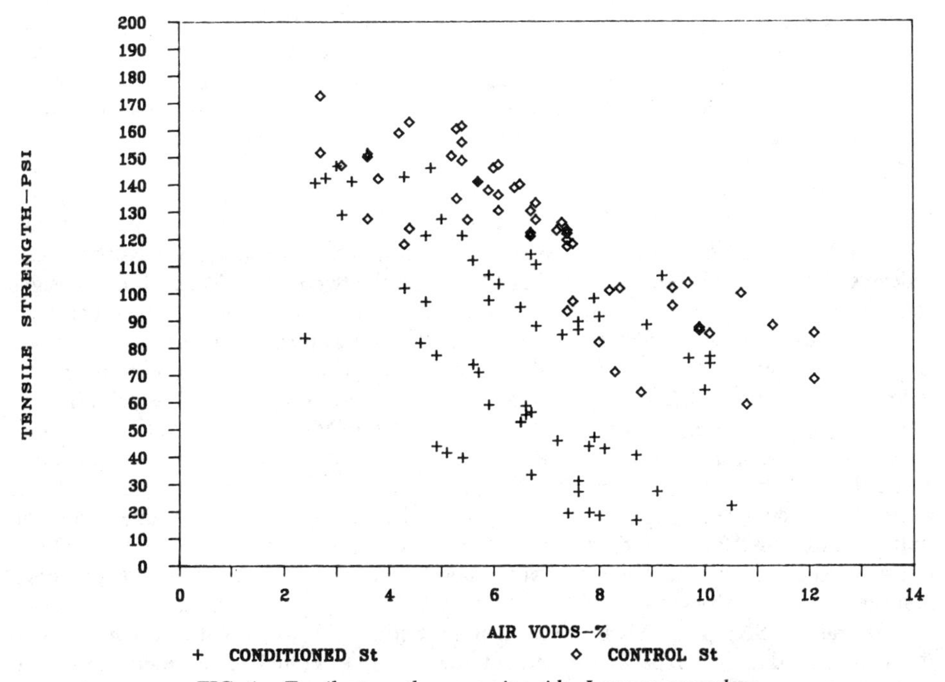

FIG. 1—*Tensile strength versus air voids, Lottman procedure.*

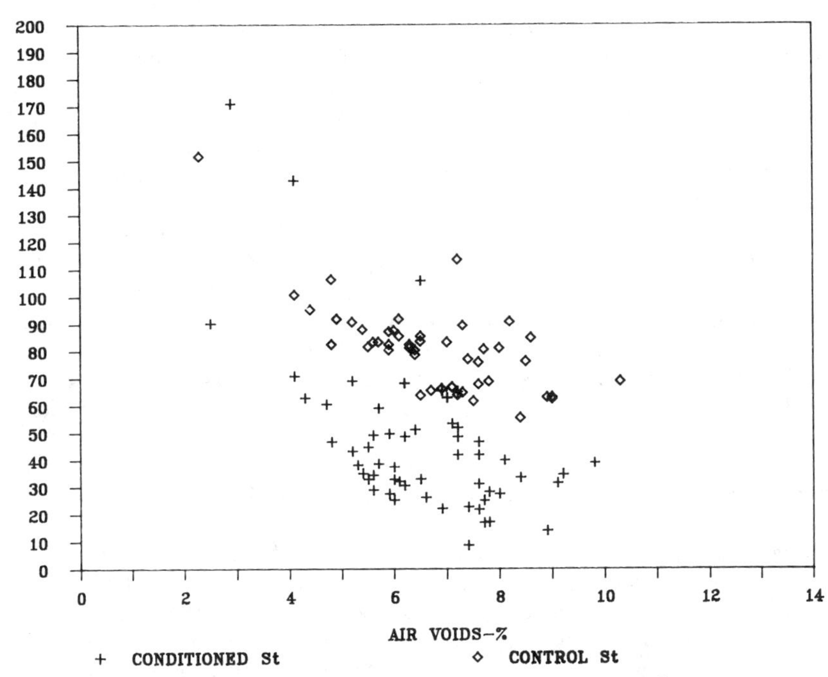

FIG. 2—*Tensile strength versus air voids, modified Lottman procedure.*

These minimum TSRs are based on the average tensile strength value of multiple samples that fall within a specified air void range. As the specimens are made they are grouped by air void content. The groups are then divided so half are used as controls and the other half are conditioned. The assumption is that since specimens with similar air void contents are in both groups, the effect of air voids is cancelled.

During normal specimen preparation an air void variation of approximately ±1% is typical for most laboratories [5,6]. Plots of the tensile strength (S_t) versus air void content indicated that the slope of the control curve was different from that of the conditioned curve. This difference in slopes caused the TSR to be more variable than expected. Differences in the average air void content of the control and conditioned specimens as little as 0.5% would cause TSR values to vary by as much as 20%.

Laboratory Procedure

Specimens were prepared from oven-dried aggregate sieved into fractions are recombined according to a given job-mix gradation. The aggregates for each specimen were weighed into aluminum pans. When hydrated lime was used, the lime was dry-mixed into the heated aggregate immediately before the heated asphalt cement was added and mixed into the sample. The asphalt cement was heated in metal containers in a separate oven. Liquid antistrip was weighed into the asphalt containers and dispersed using a paint-mixing attachment on a hand-held electric drill immediately before mixing with the heated aggregate. Asphalts were heated to equiviscosity (170 ± 20 cS_t) temperature for mixing determined by grade and source. Mixing was accomplished by a Cox mechanical mixer. Compaction was accomplished by using a Rainhart mechanical Marshall compactor. The number of blows per face was varied to achieve the desired air void content. Typically for 7 ± 1% air voids 20 blows per face were administered to each pill. As the work progressed, the number of specimens produced for each test was standardized at six. Two specimens were compacted at the Marshall design level of 50 or 75 blows. Two specimens were compacted at 20 blows and two more at 12 blows per face. This typically gave air voids ranging from 3 to 9%.

Specimens for conditioning were vacuum saturated in both procedures. Those saturated for the Lottman procedure were weighed and vacuum-saturated for 30 min using 660 mm of mercury. The vacuum was released and the specimens were allowed to soak undisturbed for an additional 30 min (Standard Operating Procedure Method GHD-66, Georgia Department of Transportation) [*3*]. Each specimen was then placed in a plastic bag with 10 cm^3 of additional water and frozen for a minimum of 15 h. After the freezing period, the samples while still sealed in plastic bags were placed in a 60 ± 2°C bath for 30 min. The plastic bags were then cut open and the specimens were allowed to remain undisturbed in the 60°C bath for 24 h. The specimens were then removed and allowed to cool to ambient temperature for 1 h.

The control (unconditioned) specimens were stored dry at approximately 25°C until tested. Both the control and conditioned specimens were cooled in a 12.8 ± 2°C refrigerator for 3 h.

Immediately after removal from the refrigerator, each specimen was tested in an indirect tensile strength head loaded at 1.65 mm/min. The maximum load obtained was recorded.

Those specimens to be conditioned by the modified Lottman procedure (Standard Method of Test 64-428-85, Kentucky Department of Transportation) [*4*] were weighed and vacuum saturated at 660 mm of mercury for 45 to 60 s. This typically produced saturation levels between 60 and 80%. No freeze-thaw cycles were used. The partially saturated specimens were conditioned by soaking in a 60°C water bath for 24 h. The conditioned specimens were then cooled in a 25°C bath for 1 h. The control specimens were placed in a 25°C water bath for 20 min prior to testing.

The maximum diametral load was then determined for both sets of specimens using an automatic recording Marshall device and an indirect tensile strength loading head.

The tensile strength of each specimen, for either procedure, was determined as

$$S_t = \frac{2P}{tD\pi}$$

where

S_t = tensile strength, Pa,
P = maximum load, kN,
t = specimen thickness, mm, and
D = specimen diameter, mm

Test Results

The test results are presented graphically in Figs. 3 to 13. These figures illustrate the strong relationship between S_t and air void content. The figures also demonstrate that for different combinations of mix type, aggregate, antistrip, and procedure the slope of the curves differ for both the control and conditioned test results. This slope difference is shown in Figs. 1, 10, and 13. The consequence of these different slopes is that the TSR values calculated by averaging the S_t values and those determined by curve fitting differ. The relationship between S_t and air voids was modeled by regression analysis. The models are presented in Table 2. These models were used to determine the corresponding S_t for the

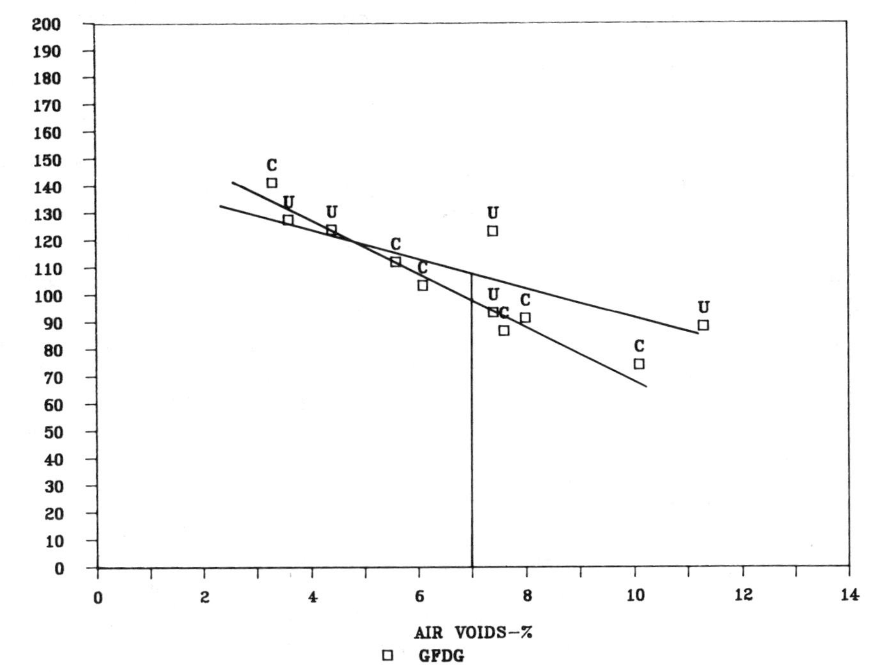

FIG. 3—*Tensile strength versus air voids, moisture sensitivity study.*

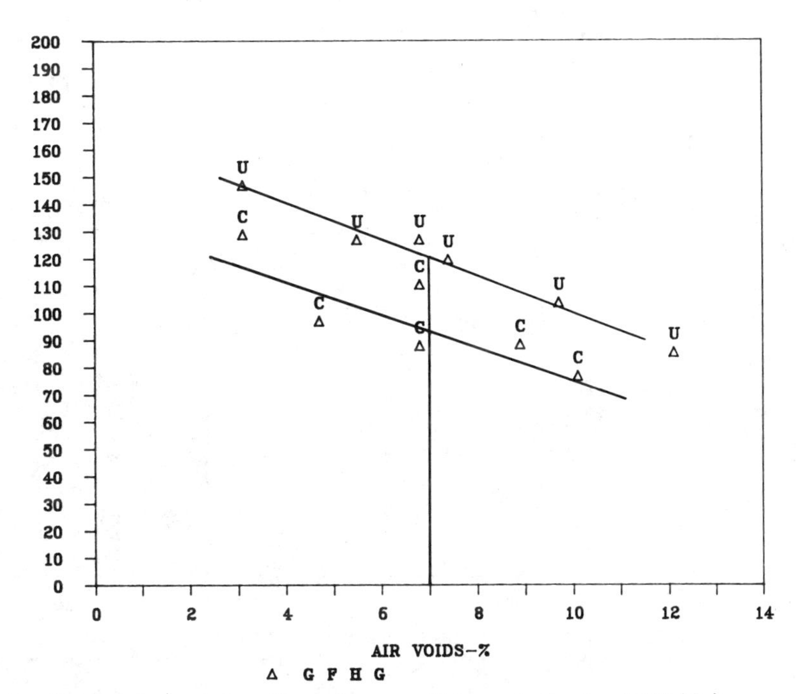

FIG. 4—*Tensile strength versus air voids, moisture sensitivity study (1 psi = 6.895 kPa).*

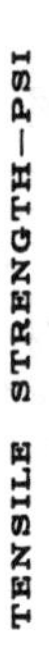

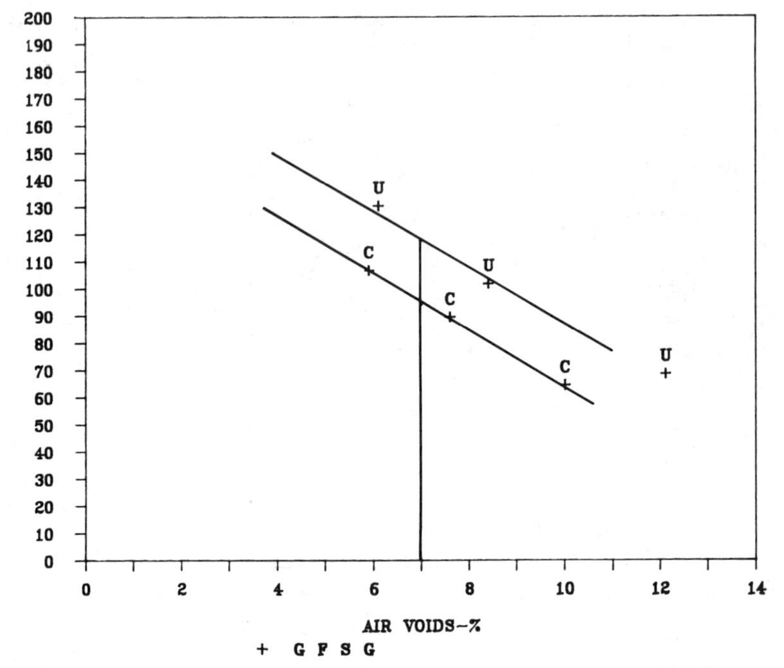

FIG. 5—*Tensile strength versus air voids, moisture sensitivity study (1 psi = 6.895 kPa).*

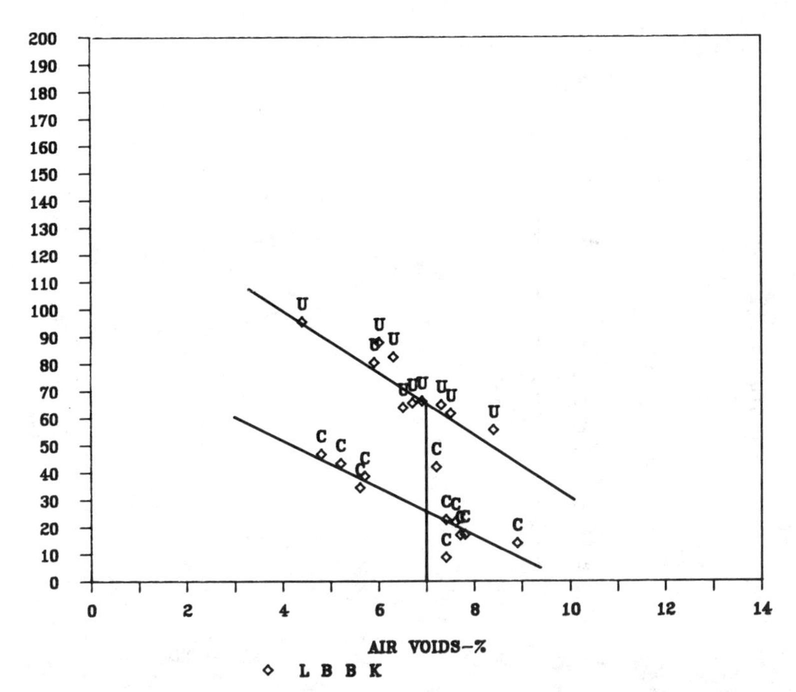

FIG. 6—*Tensile strength versus air voids, moisture sensitivity study (1 psi = 6.895 kPa).*

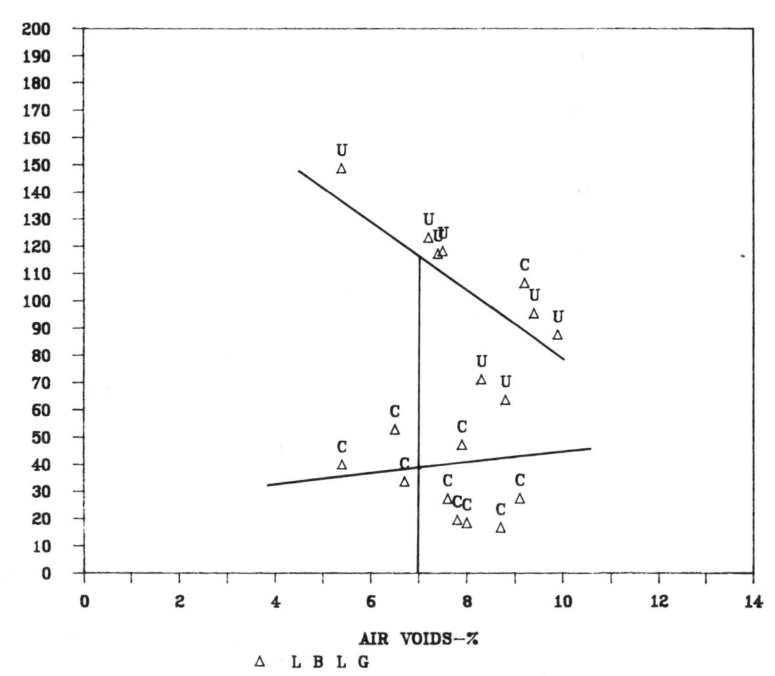

FIG. 7—*Tensile strength versus air voids, moisture sensitivity study (1 psi = 6.895 kPa).*

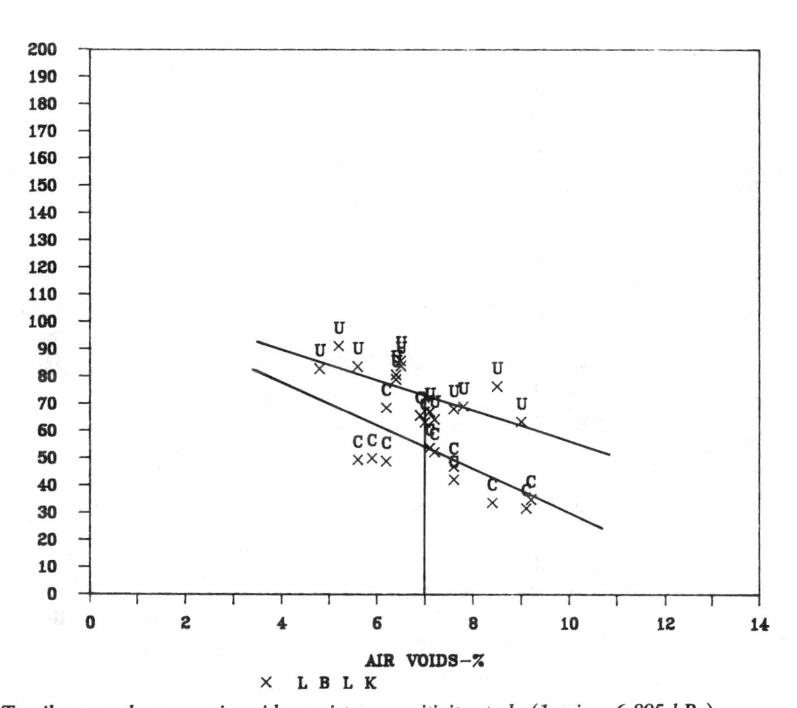

FIG. 8—*Tensile strength versus air voids, moisture sensitivity study (1 psi = 6.895 kPa).*

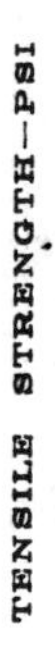

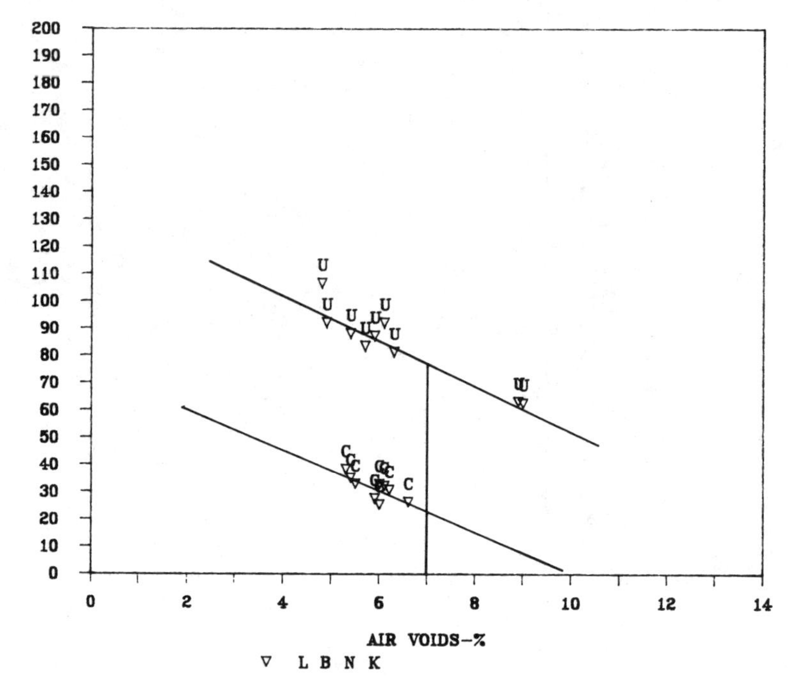

FIG. 9—*Tensile strength versus air voids, moisture sensitivity study (1 psi = 6.895 kPa).*

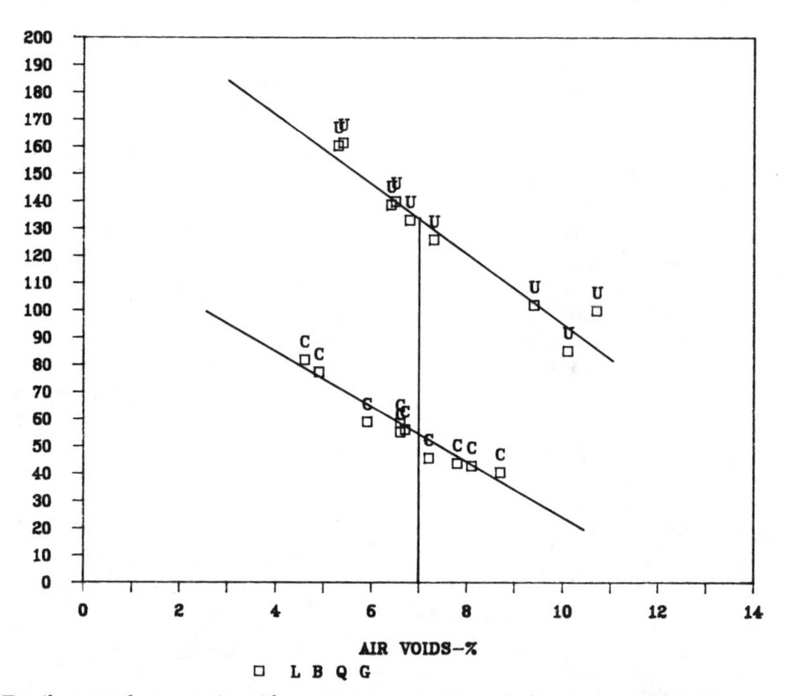

FIG. 10—*Tensile strength versus air voids, moisture sensitivity study (1 psi = 6.895 kPa).*

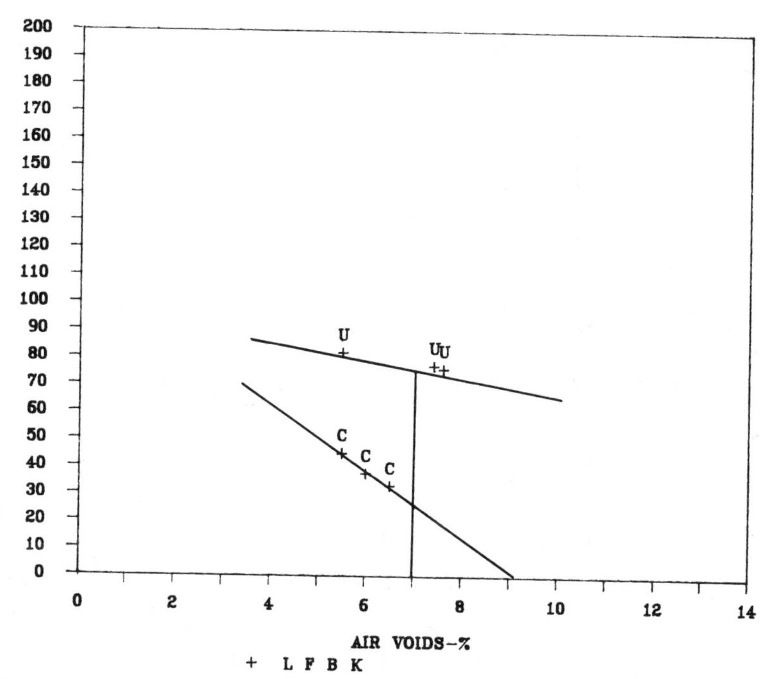

FIG. 11—*Tensile strength versus air voids, moisture sensitivity study (1 psi = 6.895 kPa).*

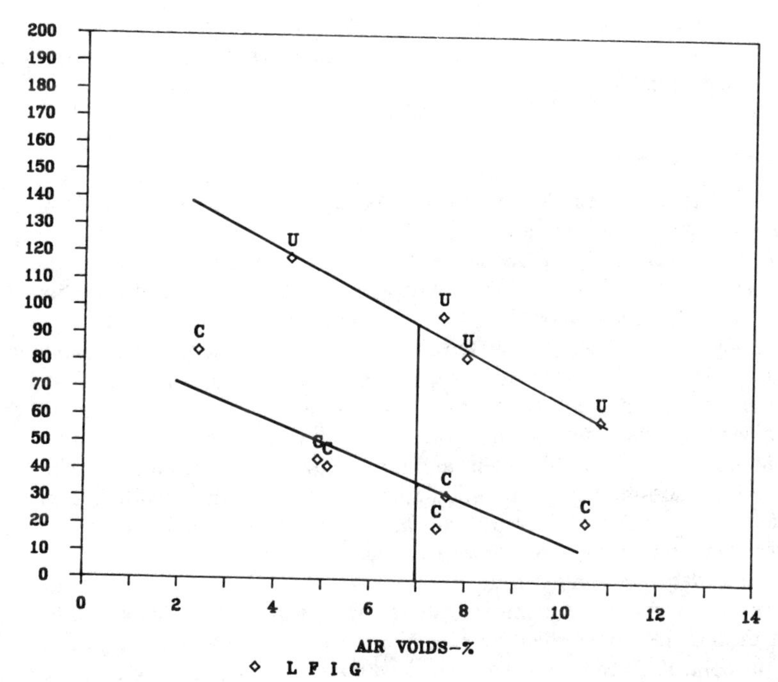

FIG. 12—*Tensile strength versus air voids, moisture sensitivity study (1 psi = 6.895 kPa).*

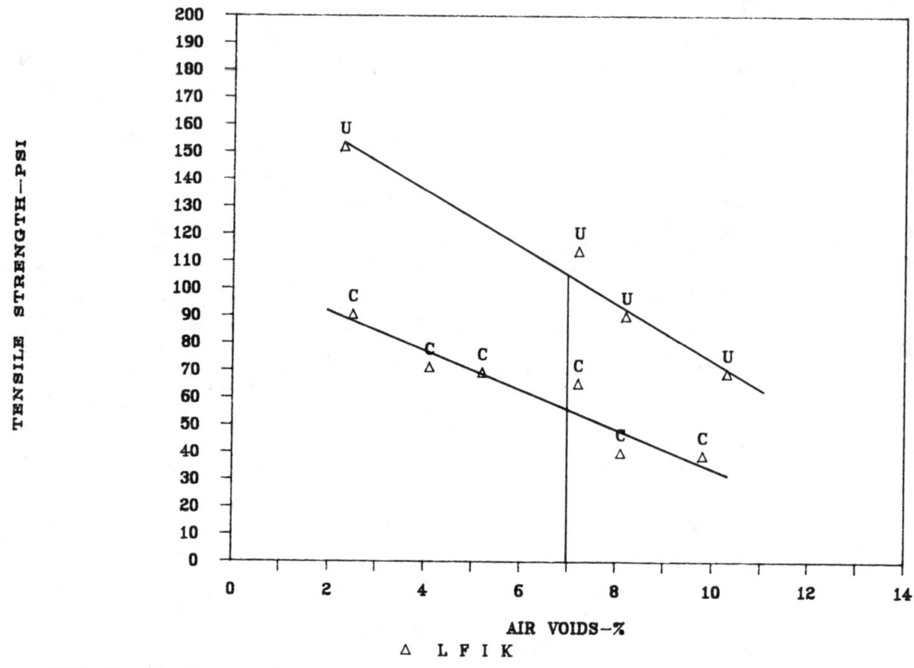

FIG. 13—*Tensile strength versus air voids, moisture sensitivity study (1 psi = 6.895 kPa).*

control and conditioned samples at the 7% air void level. The curve-fit TSR values were calculated from the predicted S_t values. A comparison of average TSR and curve TSR values is presented in Table 3.

Discussion of Results

The data presented illustrate three possible conditions: First, the control tensile strength curves are the same for treated (mixes with an asphalt modifier) and untreated mixes; second, the treated control S_t is much higher than that of the untreated control due to the addition of the asphalt modifier; and third, the slopes for the control and conditioned data vary for the various mixes.

In Fig. 14 there are three curves representing a limestone with no additive, a limestone with hydrated lime as an additive, and a limestone with 0.5% liquid antistrip. For this aggregate-asphalt combination the unconditioned specimens with and without additives achieved the same tensile strength. This is illustrated by the top curve in Fig. 14. The conditioned untreated specimens and the specimens treated with 0.5% liquid antistrip produced the bottom curve. The middle curve represents the conditioned lime-treated specimens. Looking at these curves, one can see that, depending on the air void content of the specimens, the tensile strength varies greatly.

These data are summarized in Tables 4 and 5. Demonstrated in these tables is that, regardless of mix constituents, tensile strength drops as air voids increase. The conditioned S_t's drop more rapidly than the control S_t's, in one case decreasing over 2.5 times as much as the control. Because the TSR varies with air void content, mixes should be compared by TSRs calculated at a constant air void level. To illustrate this point, at 7% air voids, Mix

TABLE 2—*Regression models for tensile strength data.*[a]

CA	Type	AS	Pro	Condition	b_0[b]	b_1[c]	S_t	TSR	r^2[d]
G	F	D	GA	c	168.2	−9.8	99.3		
n[e] = 12				u	146.3	−5.4	108.4	91.6	84.9
G	F	H	GA	c	136.4	−6.2	93.3		
n = 15				u	168.0	−6.7	121.2	77.0	87.1
G	F	S	GA	c	167.8	−10.3	95.5		
n = 6				u	190.6	−10.2	119.2	80.1	99.5
L	B	B	KY	c	86.9	−8.6	26.6		
n = 21				u	145.3	−11.1	67.9	39.2	94.0
L	B	L	GA	c	24.5	2.1	39.0		
n = 20				u	205.2	−13.0	113.9	34.2	72.8
L	B	L	KY	c	109.9	−8.1	53.2		
n = 32				u	112.9	−5.6	73.9	72.1	77.0
L	B	N	KY	c	76.4	−7.7	22.7		
n = 18				u	136.8	−8.3	78.5	29.0	98.1
L	B	Q	GA	c	125.8	−10.4	53.2		
n = 19				u	224.5	−12.8	134.5	39.5	98.5
L	F	B	KY	c	110.4	−12.0	26.4		
n = 6				u	96.3	−2.6	77.8	34.0	99.9
L	F	I	GA	c	86.7	−7.3	35.3		
n = 10				u	159.2	−9.1	95.0	37.1	93.2
L	F	I	KY	c	104.9	−6.9	56.6		
n = 10				u	178.3	−10.3	106.1	53.3	96.6

[a] Model: $S_t = b_0 + b_1$ (% air voids).
[b] b_0 = Intercept.
[c] b_1 = Slope.
[d] r^2 = Variance.
[e] n = Number of samples.

LBLK has a control S_t of 510 kPa (73.9 psi) and a conditioned S_t of 367 kPa (53.2 psi) for a TSR of 72.1% which would pass, based on a 70% minimum TSR binder mix requirement. For this same mix, if the control air void was 6% [S_t = 548 kPa (79.5 psi)] and the conditioned air voids are 8% [S_t = 311 kPa (45.1 psi)] the resulting TSR would be only 56.7%, causing the mix to fail the minimum requirements. Note, however, that both air void values are within test specifications of 7 ± 1%. If the relationship is reversed, that is, the control is taken at 8% and the conditioned at 6%, the TSR jumps up to 90%.

The probability of an undesirable mix passing the test is just as high as good mix failing, as was just discussed. At 7% air voids for both control and conditioned S_t's, mix LBBK's TSR of 39.2 substantially fails a 60% minimum TSR. If the air void levels are not matched and the control S_t is taken at 8% average air voids [S_t = 392 kPa (56.8 psi)] and the conditioned S_t is taken at 6% air voids [S_t = 243 kPa (35.2 psi)]. The new TSR is 62%, which would pass the specification requirements.

TABLE 3—*Comparison of tensile strength* (S_t) *data determined by curve fitting and averaging.*

CA	Type	AS	Pro	Condition	Calculated S_t	Measured S_t (TSR)	Calculated TSR	Average TSR	Difference, %
G	F	D	GA	c	9.3	93.8 (7.2)[c]			
n[b] = 12				u	108.4	103.4 (7.6)	91.6	90.7	−1.0
G	F	H	GA	c	93.3	91.7 (6.8)			
n = 15				u	121.2	123.4 (7.1)	77.0	74.3	−3.5
G	F	S	GA	c	95.5	89.6 (7.6)			
n = 6				u	119.2	130.4 (6.1)	80.1	68.7	−14.2
L	B	B	KY	c	26.6	21.6 (7.5)			
n = 21				u	67.9	70.4 (6.7)	39.2	30.7	−21.7
L	B	L	GA	c	39.0	35.9 (7.3)			
n = 20				u	113.9	119.5 (7.4)	34.2	30.0	−12.3
L	B	L	KY	c	53.2	57.8 (6.9)			
n = 32				u	73.9	72.7 (7.0)	72.1	79.5	+10.3
L	B	N	KY	c	22.7	29.4 (6.2)			
n = 18				u	78.5	86.7 (6.2)	29.0	33.9	+16.9
L	B	Q	GA	c	53.2	52.0 (7.0)			
n = 19				u	134.5	134.5 (6.8)	39.5	38.7	−2.0
L	F	V	KY	c	26.4	35.2 (6.3)			
n = 6				u	77.8	76.5 (7.5)	34.0	46.0	+35.3
L	F	I	GA	c	35.3	25.3 (7.5)			
n = 10				u	95.0	89.6 (7.8)	37.1	29.2	−21.3
L	F	I	KY	c	56.6				
n = 10				u	106.0	113.8 (7.2)	53.3	57.3	+7.5

[a] c = Conditioned specimens and u = unconditioned (control) specimens.
[b] n = Number of specimens tested.
[c] Average percentage air voids.

These cases illustrate an undesirable condition of too much variability in TSRs resulting from mismatching S_t's taken from dissimilar air void levels. S_t is very sensitive to air void level and thus requires that a constant void level be used for calculations. Taking the "extremes" shown above, in the first case there was a TSR range of ±16.7 percentage points, in the second case, the range was ±19.6 percentage points.

To eliminate that discrepancy in the results, a line was best-fit using both statistical techniques and by curve-fitting of the data points. Using this procedure, the curves are entered at the designated air voids content of 7% and the unconditioned tensile strength and the corresponding conditioned tensile strength are found. Using Student's T-Joint probability the range on the predicted TSR value is ±4% at a 95% confidence interval. We believe that a ±4% variation is more acceptable than the >16% variation possible with simple averaging of the S_t data.

Another series of tests were performed using limestone from a different section of the same quarry. For this particular set of results, the unconditioned tensile strength was in-

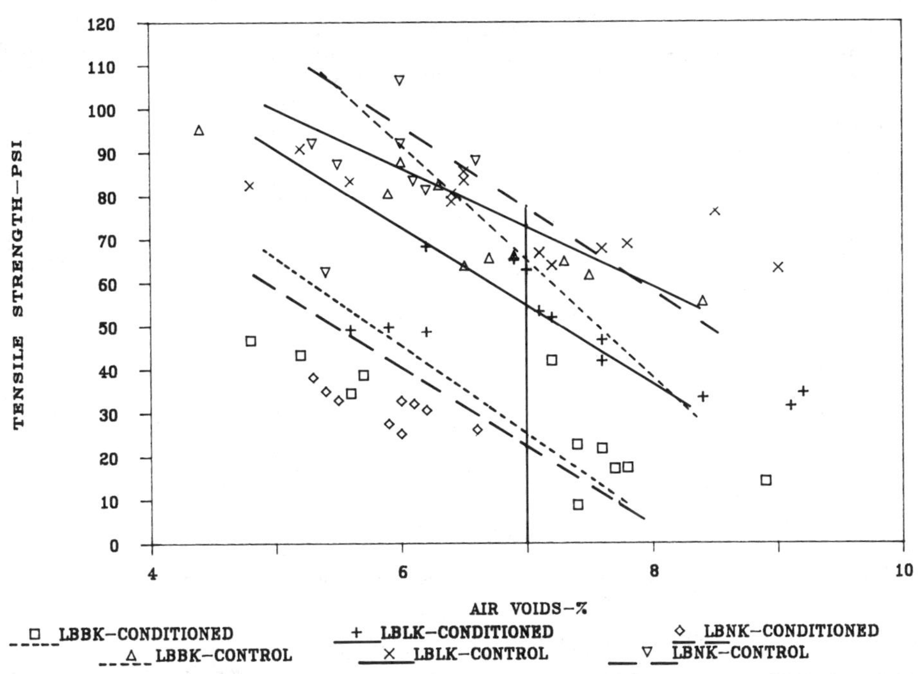

FIG. 14—*Tensile strength versus air voids, moisture sensitivity study—all data (1 psi = 6.895 kPa).*

creased with the addition of either liquid antistrip or hydrated lime from the 503 kPa (73 psi) that was obtained for the untreated aggregate up to 889 kPa (129 psi). This value was obtained for both the liquid antistrip and the hydrated lime-modified mixes. After conditioning, both asphalt modifiers gave the same conditioned tensile strength of 370 kPa (53.6 psi) at 7% air voids. This gave a TSR of 42 ± 3%. Interestingly, the addition of the modifiers to the asphalt increases the conditioned tensile strengths almost to that obtained for the untreated, unconditioned aggregate mixes. Unless the results are related back to the designated air voids content of 7%, there is a wide range of TSR results that are obtainable, as one can see in Table 5.

TABLE 4—*Effect of air voids on tensile strength* (S_t).

Mix Designation	Air Voids	Control S_t	Conditioned S_t
LBBK	6	79.0	35.2
	7	67.6	26.7
	8	56.8	18.0
LBLK	6	79.5	61.3
	7	73.9	53.2
	8	68.3	45.1
LBNK	6	86.8	30.4
	7	78.5	22.7
	8	70.2	15.0

TABLE 5—*Effect of air voids on TSR.*

Mix Designation	Percent Air Voids		TSR
	Control	Conditioned	
LBBK	6	6	44.6
	7	7	39.2
	8	8	31.7
	6	8	22.8
	8	6	62.0
LBLK	6	6	77.1
	7	7	72.1
	8	8	66.8
	6	8	56.7
	8	6	90.0
LBNK	6	6	35.0
	7	7	29.0
	8	8	21.4
	6	8	17.3
	8	6	43.3

TABLE 6—*Boil-soak test results.*

Aggregate	Size	Antistrip	Coating Retained,[a] %
L	No. 67	1% hydrated lime	32
		0.5% A	78
		0.5% B	98
		0.5% C	79
	screenings	hydrated lime	60
		0.5% A	71
		0.5% B	81
		0.5% C	79
LC	No. 67	hydrated lime	45
		0.5% A	75
		0.5% B	81
		0.5% C	81
	screenings	hydrated lime	61
		0.5% A	68
		0.5% B	81
		0.5% C	81
LD	No. 67	hydrated lime	42
		0.5% A	63
		0.5% B	99
		0.5% C	81
	screenings	hydrated lime	73
		0.5% A	72
		0.5% B	92
		0.5% C	77

[a] Based on three individual observations.

As part of the study, a boiling test was conducted to rate the different antistrips. Table 6 shows the results of using No. 67 coarse aggregate and screenings from three different limestone sources. In all cases, liquid antistrip B gave the highest percent retained coating after the boiling test. One percent hydrated lime in all cases, except for the screenings from the third aggregate source, resulted in the poorest percentage of coating retained. This verifies previous reports that the boiling test is biased toward liquid antistrips [*2,4,5*]. Figures 12 and 13 compared with Figs. 10 and 11 illustrate that 1% hydrated lime performed as well as or better than the liquid antistrips. The results of the boiling test indicate that the hydrated lime specimens would have produced the worst results.

There was some concern that the degree of saturation in the specimens would have an effect on the tensile strength and hence, the tensile strength ratio. Figure 15 is a plot of tensile strength versus percent saturation. The open circles with crosses are for specimens containing less than 6.8% air voids, the solid circles have 6.8 to 7.3% air voids, the crosses have greater than 7.3% air voids, and the solid squares at the bottom are untreated specimens. The numbers next to some of the data points indicate the actual air voids content. From these data no conclusions can be drawn on the effect of saturation on tensile strength for the limestone mixes tested.

Another concern was with the effect of the freezing cycle on the tensile strength ratios. A number of the 1% lime mixes were tested using both the Lottman and modified Lottman procedures. These results are presented in Table 7. These results indicate that the Lottman procedure is more severe than the modified Lottman procedure. Comparison of identical mixes shows in all cases the Lottman test produced TSRs lower than did the modified Lottman. Whether by curve fitting or data averaging, Mix LBL, a binder mix, had Lottman results that were less than 50% of the corresponding modified Lottman. For Mix LFI, a surface mix, the trend was the same, with the difference not quite as dramatic (30%). These

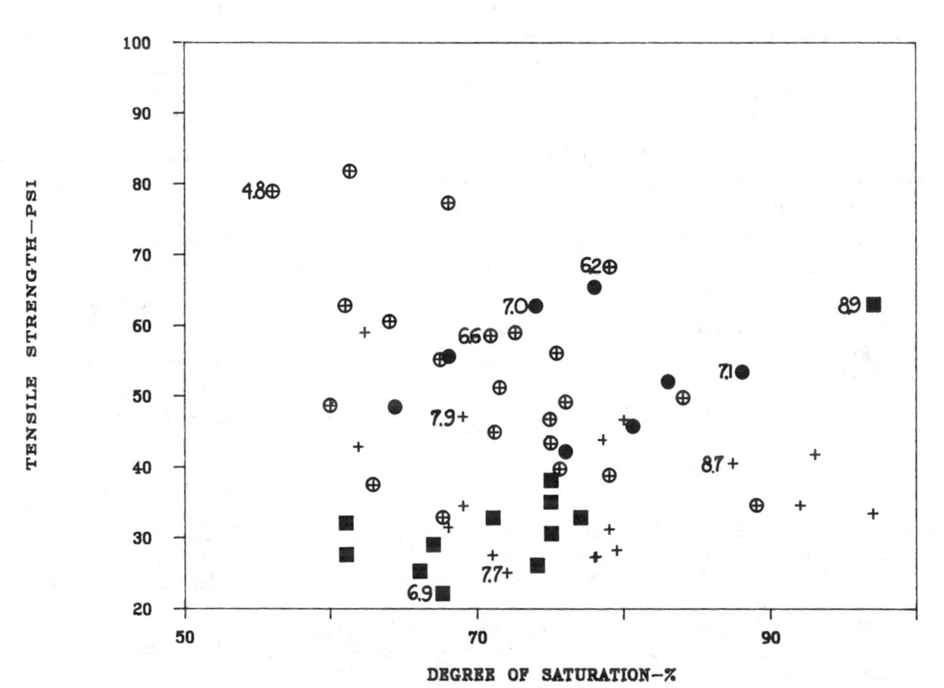

FIG. 15—*Tensile strength versus saturation, moisture sensitivity study (1 psi = 6.895 kPa).*

TABLE 7—*Comparison between Lottman and modified Lottman procedures.*

Specimen	Lottman	Modified Lottman	TSR Difference, %	Sign. at a = 0.05
		CURVE FITTED VALUES		
LBL	34.2	72.1	37.9	yes
LFI	37.1	53.3	16.2	yes
		AVERAGED VALUES		
LBL	30.1	79.5	49.5	yes
LFI	28.2	57.3	29.1	yes

results indicate that regardless of particle size the Lottman procedure is more severe than the modified procedure [*4,6*].

Conclusions

1. The tensile strength ratio should be determined at 7% air void levels that are determined statistically or graphically. The advantage of this is that it reduces the test variation and makes it very easy for the people in the laboratory to plot the results from all the pills produced and best-fit a curve through the data. This has the benefit of reducing the number of specimens needed for testing.

2. The specifications should be two-tiered: along with the minimum tensile strength ratio, a minimum after conditioning tensile strength should be specified. One of the results presented indicated that the tensile strength obtained after conditioning a treated mix was higher than that obtained for the unconditioned, untreated mix. However, due to the increase in strength of the unconditioned mix with the additive, the tensile strength ratio did not increase enough to meet a minimum TSR requirement. We feel that if the aggregate-asphalt modifier combination has a tensile strength higher than the untreated aggregate, this improvement should be taken into account.

3. The tensile strength ratio is a function of air voids. Results in this study do not show that saturation greatly affects the tensile strength ratio. However, that result may be due to the small range of saturation obtained in this study.

4. The Lottman procedure is more severe than the Tunnicliff-Root procedure. The final question is what value of unconditioned strength should be used to determine the tensile strength ratio. Using the unconditioned, unmodified tensile strength as a divisor instead of the unconditioned modified tensile strength substantially changes the tensile strength ratio. This method may remove the effects of the S_t improvements of the unconditioned modified mixes. Using the TSR determined in this fashion along with a minimum tensile strength value may give a more reasonable indication of how a mix will perform under field conditions.

References

[*1*] Kennedy T. W., Roberts, F. L., and McGennis, R. B., "Effects of Compaction and Effort on the Engineering Properties of Asphalt Concrete Mixes," *Placement and Compaction of Asphalt Mixes, ASTM STP 829,* F. T. Wagner, Ed., American Society for Testing and Materials, Philadelphia, 1984, pp. 48–66.

[*2*] Busching, H. W., "A Statewide Program to Identify and Prevent Stripping Damage," presented at the American Society for Testing and Materials Symposium on Water Damage of Asphalt Pavements, Williamsburg, VA, 12 Dec. 1984.

[3] Lottman, R. P., "Predicting Moisture-Induced Damage to Asphalt Concrete," National Cooperative Highway Research Program *Report 192,* Transportation Research Board, 1978.
[4] Tunnicliff, D. G. and Root, R. E., "Use of Antistripping Additives in Asphaltic Concrete Mixtures Laboratory Phase," National Research Program *Report 274,* Transportation Research Board, 1984.
[5] O'Connor, D. L., "Action Taken by the Texas Department of Highways and Public Transportation to Identify and Reduce the Stripping Potential of Asphalt Mixes," *Proceedings,* Association of Asphalt Paving Technologists, Vol. 53, 1984, pp. 631–638.
[6] Tunnicliff, D. G. and Root, R. E., "Testing Asphalt Concrete for Effectiveness of Antistripping Additives," *Proceedings,* Association of Asphalt Paving Technologists, Vol. 52, 1983.

DISCUSSION

William Ciggelakis[1] (*written discussion*)—The tensile strength ratio (TSR) method of evaluation of a mix design may not be considered as the single common procedure. While this test method is a good general procedure to evaluate moisture susceptibility, it is not without deficiencies. Our testing of a granite aggregate revealed an inconsistency with respect to improvement of mix characteristics.

The Texas Highway Department and Public Transportation Method 531-C was observed as the test procedure. Our testing revealed a reduced indirect tensile strength of the dry specimen with antistripping agent in comparison to the dry specimen without an antistripping agent. There was no appreciable difference of indirect tensile strength between the conditioned specimens with or without the antistripping agent. However, an improved TSR ratio was observed. The improved TSR ratio was in excess of 70%, and the mix may be considered for placement.

A concern lies with the fact that material strength properties may be reduced and the material may be considered acceptable. However, the same mix when tested to be tougher or to have greater indirect tensile strength may be disqualified. The TSR currently does not address prescribed minimum values that could be used in the denominator to treat different characteristics equally in evaluating the performance of a mix.

Our testing has not been of a magnitude to be conclusive in recommending a minimum test value. The establishment of such criteria would be a welcome addition to the prescribed TSR and would treat performance more fairly.

Ervin L. Dukatz, Jr., and Richard S. Phillips (*authors' closure*)—The authors agree with Mr. Ciggelakis' comments. As we discussed in the paper, we have similar concerns. We recommend that the *engineer* in charge look at the wet and dry tensile strength values for the modified (treated) and unmodified mixes. A modified mix with a higher wet tensile strength than the dry tensile strength of the unmodified mix will perform better, regardless of the TSR value. It is because of these problems that we do not advocate a "cook-book" analysis of the data.

[1] McClelland Engineers, Inc., 9750 Tanner Road, Houston, TX 77041.

Charles S. Hughes[1] *and G. William Maupin, Jr.*[2]

Factors that Influence Moisture Damage in Asphaltic Pavements

REFERENCE: Hughes, C. S. and Maupin, G. W., Jr., **"Factors that Influence Moisture Damage in Asphaltic Pavements,"** *Implication of Aggregates in the Design, Construction, and Performance of Flexible Pavements, ASTM STP 1016,* H. G. Schreuders and C. R. Marek, Eds., American Society for Testing and Materials, Philadelphia, 1989, pp. 96–102.

ABSTRACT: This paper addresses the increase in moisture damage occurring in asphalt pavements over the past decade. It defines moisture damage and discusses the many causal factors that have been cited, with emphasis on those under the control of the asphalt technologist. The increase in heavy traffic loadings is mentioned as the number one cause of damage, and the potential for poor selection and handling of aggregates to exacerbate the problem is noted. Several methods of evaluating asphalt mixes for susceptibility to moisture damage are given, and the need for a predictive procedure approved by a nationally recognized authority is emphasized. Lastly, several methods of reducing the potential for moisture damage are discussed.

KEY WORDS: additive, adhesion failure, aggregate, asphalt mix, cohesion failure, laboratory test, moisture damage, moisture susceptibility, prediction, slurry seal, stripping, tensile strength ratio (TSR), traffic volume, voids in mineral aggregate (VMA)

It is apparent that moisture damage in asphalt pavements has increased over the last several years, particularly in the southeastern part of the United States, and there has been much speculation over the causal factors [*1–3*]. The major factors cited relate to the asphalt cements and aggregates being used, surface type, pavement age, and traffic volumes and weights.

Definition of Moisture Damage

Although moisture damage has been recognized as a mode of pavement distress for years [*4*], there seem to be many different ideas as to when and how it occurs. It is evidenced by either adhesion failure, the most prevalent form, or cohesion failure, or possibly a combination of the two. An adhesion failure occurs as a complete or partial separation of the asphalt film from the aggregate in the presence of moisture. Some technologists hold the opinion that a stain or discoloration of the aggregate left by the separation of the asphalt film does not constitute "stripping." In the authors' opinion, this separation constitutes moisture damage and thus stripping, irrespective of the number of asphalt molecules that remain on the aggregate surface. Other opinions attach importance to oxidation of the asphalt film as a contributor to moisture damage. Again, in the authors' opinion, if the

[1] Senior research scientist, Virginia Transportation Research Council, Charlottesville, VA 22403-0817; formerly, technical representative, Carstab Products, Cincinnati, OH 45215.

[2] Research scientist, Virginia Highway and Transportation Research Council, Charlottesville, VA 22903-0817.

separation of the asphalt and aggregate takes place in the presence of moisture, moisture damage occurs, irrespective of the age or hardness of the asphalt film.

A cohesion failure occurs as a separation within the asphalt film in the presence of moisture. This failure is harder to visualize than the adhesion type and can be confused with other failure mechanisms that cause a rupture of the asphalt film, which possibly accounts in part for this failure mechanism not being reported more often.

How much moisture is necessary to cause moisture damage? This is an intriguing question. In Virginia it has been noted that moisture damage often has appeared after a period of rainy weather, which might lead one to think that a reasonably high amount of moisture is necessary to cause damage. However, when some of the damaged pavements have been removed for rehabilitation, no sign of free moisture has been evident in any of the asphaltic layers or in the shoulder. And this has led one to ask, Did the pavement dry out completely? or, Is moisture vapor pressure alone sufficient to cause or continue moisture damage? Corollary questions are, How much healing of moisture damage takes place in hot, dry weather? and, Does the vapor pressure affect healing?

These questions are raised to illuminate the fact that although moisture damage has been observed for over 50 years, some of the basic mechanism are not well understood. As will be discussed, solutions to the problem are being sought with little insight into the basic mechanisms of the failure.

Factors Affecting Moisture Damage

As indicated earlier, many factors affect moisture damage. At least two of these are beyond the control of the engineer. It is generally agreed that traffic volume and weight play an important part in moisture damage. Mixes that have performed well for several years under low to moderate levels of traffic can show moisture damage quickly under high levels. The line between an acceptable and unacceptable traffic level for a particular mix is usually determined by experience, and for convenience is usually applied to a road category; for example, some aggregates may be banned from use in pavements on the interstate system. In the opinion of the authors, the most important factor in the more extensive occurrence of moisture damage is the increase in traffic, specifically the increases in the weight and number of loads and in tire pressures. By far the greatest increase in moisture damage has taken place on interstate highways, where the increase in high-speed, heavy loads has been greatest. The fact that the number of tractor trailers and buses on Virginia interstate highways increased 9% in 1985 indicates that the problem will tend to get worse if corrective measures are not taken.[3] Because the relation between accumulated traffic loadings and pavement deterioration is usually not linear, a moderate increase in traffic could produce a drastic increase in moisture damage.

The age of the paving mix is another factor in moisture damage, but is beyond the control of the engineer. The best the engineer can do is to control the effects of aging by reducing the voids in the mix to a level sufficiently low to retard oxidation of the asphalt film. Of course, an added advantage of this solution is that it reduces the ability of water to permeate the mix.

The factors that the engineer has at least some control over are the types and condition of the aggregate and asphalt to be used, the combination of these two, and, most important, the mix design. Because the increase in moisture damage was approximately concomitant with the Organization of Petroleum Exporting Countries (OPEC) oil embargo, many paving

[3] McGhee, K. H., personal communication concerning traffic increase on Virginia's interstate highway system, July 1986.

technologists have attributed the problem to changes in the asphalt cements used and the general decrease in their affinity for aggregates. This attribution has not been substantiated by tests on the properties of the asphalt cements, although subjectively it appears to have some basis. Some technologists have cited the increased use of low-quality aggregates brought about by the increased road-building program that has been under way since the late 1950s. This latter reasoning is vitiated by the fact that many aggregates from the same geological formation, and indeed from the same quarry, that did not have an apparent history of susceptibility to moisture damage started showing this tendency in recent years.

By selecting the proper gradation and asphalt content, the air voids and interconnected voids can be controlled, thus often making a questionable aggregate source acceptable. The key here is to have a method of predicting the performance of the mix. It is imperative that the prediction of moisture damage be based upon tests of the mix because the moisture susceptibility of either an aggregate or asphalt from a given source may change over time. An example of this change was recently reported in Virginia, where an aggregate source that historically had a tensile strength ratio (TSR) in the modified Lottman test in the high 0.40s was found to have a TSR in the low 0.90s [*5*].

There is also evidence that the types of surfaces being placed on asphalt pavements have added to their proneness to undergo moisture damage [*6,7*]. This was brought out in the Moisture Damage Workshop sponsored by Carstab Products in 1985, which concluded that surfaces which seal in moisture would be detrimental.[4] These surfaces are open-graded friction courses and slurry seals, both of which are used to seal moisture out of the underlying pavement. In certain situations they may also tend to seal the moisture in the pavement, especially when used in a resurfacing program.

Methods of Evaluating Moisture Damage

Several laboratory methods are being used to predict moisture damage. All, by necessity, include some method of accelerating the damage in a laboratory test since the failure mechanism may take from two to ten years to actually develop on the road. Most methods accelerate the damage so that a prediction can be made in a matter of hours or, at most, a few days. Any phase of testing or conditioning that requires longer than a few days places a severe practical limitation on the evaluation procedure.

Only one laboratory method—the immersion-compression test (Test Method for Effect of Water on Cohesion of Compacted Bituminous Mixtures [ASTM D 1075])—is recognized by the American Society for Testing and Materials (ASTM). In this test, which has been around for many years, strength comparisons between conditioned and unconditioned 4 by 4-in. (10 by 10-cm) cylindrical samples are used to predict moisture susceptibility. The test is not very popular because it is time-consuming and requires equipment not ordinarily found in a bituminous mix lab. Also, a report by New York State [*8*] indicates that the test gives a poor prediction of moisture damage.

Although many versions of the "boiling test" are used in the laboratory, none has ASTM's sanction. However, the boiling water test (Test Method for Effect of Water on Bituminous-Coated Aggregate-Quick Field Test [ASTM D 3625]) is sanctioned as a quick field test. Although it is quick and simple, it is very subjective, requiring a visual estimate of moisture damage.

A modification of the immersion-compression test uses the Marshall stability test on conditioned and unconditioned specimens. Typical evaluations of mixes that show moisture damage in the field but an increased Marshall stability after conditioning do nothing to

[4] Maupin, G. W., Jr., interoffice memo on Moisture Damage Workshop, May 1985.

enhance the reputation of this test as a predictor of moisture damage, irrespective of its simplicity.

The freeze-thaw pedestal test was developed by Plancher et al. [*9*] and has been used by Kennedy [*10*]. Its disadvantages are that it takes many days (25) to indicate if a mix is resistant to moisture damage and it apparently promotes the loss of cohesion as the failure mechanism as opposed to adhesion.

In the 1970's, Lottman developed and reported on a test method through the National Cooperative Highway Research Program (NCHRP) that uses the ratio of a conditioned (saturation plus a freeze-thaw cycle) strength to an unconditioned strength tested in the indirect tensile mode [*11*]. This work used a 55°F (18°C) test temperature and a loading rate of 0.065 in./min (1.6 mm/min). To adapt this procedure for use in most testing labs, modifications of the original Lottman procedure were made to allow a 77°F (25°C) test temperature and the typical Marshall loading rate of 2 in./min (51 mm/min). Maupin, among others, has shown a good correlation between the two procedures [*3*].

More recently, again through NCHRP funding, Tunnicliff and Root [*12*] have eliminated freeze-thaw cycle from the modified Lottman concept and controlled the degree of saturation in an attempt to minimize potential damage from volume changes to produce another evaluation of moisture damage.

A variation of the Lottman procedure developed by Carstab Products uses the Lottman procedure with multiple freeze-thaw cycles [*13,14*].

In addition to the indirect tension test, resilient modulus tests have been run on the same conditioned and unconditioned specimens to produce a resilient modulus ratio. This procedure is handy because it is nondestructive and thus later can be used with an additional evaluation, such as the indirect tension test. However, it is not widely used because of the lack of equipment for measuring the resilient modulus.

If the degree of saturation is limited, which is certainly logical, the TSR evaluations between the modified Lottman and Tunnicliff and Root's methods differ essentially only in the conditioning mode. But the conditioning mode is important because of the need for a test method that accurately predicts the damage. The test method must also be simple and as quick as practical. While there are several candidate methods, there is no obviously superior one. But the need is sufficiently great that ASTM D-4 Subcommittee D04.22 must find a procedure that will predict moisture damage for mixes used in modern-day traffic. This need cannot be overemphasized.

Methods of Reducing the Potential for Moisture Damage

Mix Design

A good mix design is crucial in obtaining a mix resistant to moisture damage. The difference between a mix resistant to damage and one that is not can rest upon mix design alone. The optimum asphalt content must be determined to provide an air void content in the range of 3 to 5%. And in the interest of economy, the gradation must be selected to balance the volume of voids in mineral aggregate (VMA) to allow use of a reasonable asphalt content. Of course, the mix design must be conducted on the aggregate blend and asphalt cement to be used in the field.

Aggregates from different locations in a quarry may vary considerably in moisture susceptibility, and the elapsed time after crushing can influence the stripping potential of some aggregates.[5] These normal variabilities in materials may explain some instances where strip-

[5] Workshop on Moisture Damage of Asphalt Concrete Mixtures, Cincinnati, OH, 19–20 March 1985.

ping tests conducted on samples from the plant suddenly begin to give failing results for no apparent reason.

If the gradation and optimum content are not chosen properly, one may wind up with a mix that cannot be corrected with antistrip additive, liquid or dry, to reduce moisture damage.

Moisture Damage Evaluation

Just as important as the mix design is the procedure for predicting moisture damage. As discussed previously, although none of the recently developed methods using the indirect tension test have been sanctioned by a standardization body, several are very useful for discriminating between mixes (and additives) that are resistant to moisture damage and those that are not. It is imperative that a moisture damage evaluation be part of the mix design process, particularly in areas where moisture damage is prevalent. If we may digress briefly, it is possible, and even probable, that there are areas where moisture damage occurs but is not recognized. Therefore, it would be prudent for the mix design procedure used in any location to contain a proviso on moisture damage. The necessary features of the evaluation procedure, as mentioned earlier, are speed, facileness of use, and accuracy of the prediction.

As mentioned above, it is also necessary that the moisture damage evaluation be conducted using the aggregate blend and asphalt cement to be used in the field with the job mix formula determined from the mix design. However, the compaction level must be modified to provide an air void content consistent with that often found in the field. Although at the time of construction some air void contents may range from 4 to 5%, in the authors' opinion this is rare. If this void level occurred consistently, it could dramatically improve the resistance to moisture damage by itself. More typically, the air voids found in the field are in the 6 to 8% range or, unfortunately, even higher. Therefore, it is generally agreed that in the moisture damage evaluation procedure a good target for the air void content is 6 to 8%.

The moisture damage evaluation should be used to determine:

1. Whether the mix is resistant to moisture damage or whether an antistrip additive is needed.
2. If an antistrip is needed, which one works best.
3. At what concentration [*11*].

Without these determinations, the asphalt technologist is just guessing.

Field Screening Test

There is no question that a moisture damage evaluation during the mix design stage is extremely important. And if no changes occur during production, further tests might not be necessary. However, in most asphalt concrete, production changes do take place, in aggregates as well as in asphalt cements and additives. Therefore, it is only prudent to have a quick screening test for the mix produced. This text is just an indicator of whether or not the system developed during the mix design process is still functional.

Although it is subjective, the most practical test for this purpose is the boiling water test (ASTM D 3625). It should be run fairly often (from daily to twice weekly), depending upon the variability in the moisture damage that has been experienced with the mix. A less subjective test for field use is desirable.

Field Construction

In order to reduce moisture damage, close attention to construction procedures is extremely important. And in this process, compaction is the most important factor. As stated previously, if the air void content at construction is reduced to the 4 to 5% range, moisture damage would, by and large, be eliminated. Not only would little moisture get into the mix, but interconnected voids would be reduced so that the little moisture that does get in would not be able to travel throughout the mix. And if good mix design procedures are used, the mix constructed at a 4 to 5% air void content would not overdensify and thus exacerbate other problems.

Decreasing the permeability of the pavement at more typical air void contents is another way of reducing moisture damage during construction. The use of rubber-tired rollers [*15*] has been recognized for years as a way to seal the pavement surface and reduce moisture infiltration. Unfortunately, many modern-day compaction trains do not use rubber-tired rollers.

Antistrip Additives

Antistrip additives have been recognized as being beneficial in reducing moisture damage for many years [*16*]. They come in two forms—dry, usually hydrated lime or portland cement, and liquid. While both forms have proven to be effective in certain instances, a review of the literature reveals that neither appears to be effective in all aggregate/asphalt systems. This reemphasizes the need for a moisture damage evaluation procedure.

The use of dry additives predates that of liquid antistripping additives. The latter were developed primarily as economical alternatives to the former, but are often preferred because of ease of handling.

Maupin has used the modified Lottman test to evaluate the resistance to moisture damage of mixes using lime, liquid antistrip additives, and no additives. The test results for some of these mixes placed in field test sections are shown in Table 1. Generally, the TSR appears to be an indicator of the amount of stripping in the cores.

Conclusions

Although this paper is somewhat of a state-of-the-art summary on moisture damage, several conclusions do appear to be supported:

1. Moisture damage is an increasing problem, particularly in certain areas of the country.

TABLE 1—*Tensile strength ratios (TSR).*

Route	Type of Additive	TSR	Stripping in Cores (4–5 years)
360 (B)	no additive	0.45	severe
	chemical additive No. 1	0.85	moderate
	chemical additive No. 2	0.97	slight
	lime	0.81	slight-moderate
360 (M)	no additive	0.66	moderate-severe
	chemical additive	0.89	moderate
	lime	0.95	very slight

2. An accurate method of predicting moisture damage is sorely needed at the mix design stage, and the adoption of this method by a standardizing agency is essential.

3. To minimize moisture damage, good mix design and construction techniques must be employed.

4. The moisture damage prediction test must be run on the aggregate/asphalt/additive system to be used in the field.

References

[1] Hay, R. E., "Asphalt Stripping Conference," summary prepared by chief, Materials Division, Federal Highway Administration, Atlanta, GA, 1982.

[2] Stapler, T., "A Survey of Bituminous Pavement Distress Attributable to Water Damage as Reported to the SASHTO States," survey conducted by vice-chairman, American Association of State Highway and Transportation Officials Region 2, Operating Subcommittee on Materials, 1981.

[3] Maupin, G. W., Jr., "Implementation of Stripping Test for Asphaltic Concrete," Transportation Research Board, Report No. TRR 712, Washington, DC, 1979.

[4] Nicholson, V., "Adhesion Tension in Asphalt Pavements, Its Significance and Methods Applicable in Its Determination," *Proceedings,* American Association of Paving Technologists, Vol. 3, 1932.

[5] Hughes, C. S., "Installation Report—Experimental Mixes on Richmond-Petersburg Turnpike—1985," VHTRC 86-R26, Virginia Highway and Transportation Research Council, Charlottesville, VA, Jan. 1986.

[6] Busching, H. et al., "Effects of Selected Asphalts and Antistrip Additives on Tensile Strength of Laboratory—Compacted Marshall Specimens—A Moisture Susceptibility Study," presented at the Annual Meeting of the American Association of Paving Technologists, Clearwater, FL, Feb. 1986.

[7] Maupin, G. W., Jr., "Guidelines for Rehabilitating Flexible Pavements," VHTRC 86-R29, Virginia Highway and Transportation Research Council, Charlottesville, VA, Jan. 1986.

[8] Gupta, P. K., "Evaluation of the Immersion-Compression Test for Asphalt-Stripping Potential," Special Report 77, New York State Department of Transportation, 1984.

[9] Plancher, H., Miyake, G., Venable, R. L., and Peterson, J. C., "A Simple Laboratory Test to Indicate the Susceptibility of Asphalt-Aggregate Mixtures to Moisture Damage During Repeated Freeze-Thaw Cycling," *Proceedings,* Canadian Technical Asphalt Association Meeting, Victoria, BC, 1980.

[10] Kennedy, T. W., Roberts, F. L., and Lee, K. W., "Evaluating Moisture Susceptibility of Asphalt Mixtures and the Effectiveness of Antistripping Agents Using the Texas Freeze-Thaw Pedestal Test," *Proceedings,* American Association of Paving Technologists, Vol. 51, 1982.

[11] Lottman, R. P., "Predicting Moisture-Induced Damage to Asphaltic Concrete," NCHRP Report 192, National Cooperative Highway Research Program, 1978.

[12] Tunnicliff, D. G. and Root, R. E., "Use of Antistripping Additives in Asphaltic Concrete Mixtures," NCHRP Report 274, National Cooperative Highway Research Program, Dec. 1984.

[13] Gilmore, D. W., Lottman, R. P., and Scherocman, J. A., "Use of Indirect Tension Measurements to Examine the Effect of Additives on Asphalt Concrete Durability," *Proceedings,* American Association of Paving Technologists, Vol. 53, April 1984.

[14] Scherocman, J. A., Mesch, K. A., and Proctor, J. J., "The Effect of Multiple Freeze-Thaw Cycle Conditioning of the Moisture Damage in Asphalt Concrete Mixtures," presented at the Annual Meeting of the American Association of Paving Technologists, Feb. 1986.

[15] Zube, E., "Compaction Studies of Asphalt Concrete Pavement as Related to the Water Permeability Test," RB Bulletin No. 358, Highway Research Board, Nov. 1962.

[16] Winterkorn, H. F., Eckert, G. W., and Shipley, E. B., "Testing the Adhesion Between Bitumen and Mineral Surfaces with Alkaline Solutions," *Proceedings,* American Association of Paving Technologists, Vol. 9, Dec. 1937.

Kang W. Lee[1] *and Mohamad A. Al-Dhalaan*[2]

Rutting, Asphalt Mix-Design, and Proposed Test Road in Saudi Arabia

REFERENCE: Lee, K. W. and Al-Dhalaan, M. A., "**Rutting, Asphalt Mix-Design, and Proposed Test Road in Saudi Arabia,**" *Implications of Aggregates in the Design, Construction, and Performance of Flexible Pavements, ASTM STP 1016,* H. G. Schreuders and C. R. Marek, Eds., American Society for Testing and Materials, Philadelphia, 1989, pp. 103–119.

ABSTRACT: The drastic change in traffic characteristics due to a crash development program in Saudi Arabia has introduced new forms of pavement distresses. Rutting, which was virtually absent ten years ago, became the principal distress mode. A feasibility study was performed on the Dhahran-Abqaiq road, which is experiencing a severe rutting problem, to formulate rehabilitation alternatives. The comparative study indicates that the Hveem mix-design method gives relatively lower optimum binder content than the Marshall mix-design method and that some of the potential reasons for the rutting at the Dhahran-Abqaiq road were overasphalting, fine-graded mix, and poor quality control. In addition, test results indicate that a sulfur-extended-asphalt mixture can be utilized for reducing the rutting problem, since it provided higher stability values. Finally, to help solve the widespread rutting problem in Saudi Arabia, a decision was made to include test sections in the rehabilitation program of this road.

KEY WORDS: pavements, asphalt mixtures, mix-design, rutting, sulfur-extended-asphalt, rehabilitation, hot climate, heavy load, roads

The drastic change in traffic characteristics due to accelerated development programs in Saudi Arabia has introduced new forms of pavement distress. Rutting, which was virtually absent ten years ago, has become the principal distress mode.

Rich and densely graded mixes that provided adequate stability and durability in the past have proved to be unstable under the extremely heavy axle loads of today. Consequently, new mixtures deformed and old oxidized mixtures cracked under these heavy loads.

Recent axle load surveys [*1*] indicate that the average gross load and axle loads in Saudi Arabia and other Gulf States roads exceed the maximum allowable limits in many developed countries. Tire pressures exceeding 1000 kPa (145 psi) in steel-reinforced tires are very common in the Gulf area. Such factors make the transfer of pavement technology and experience from developed countries a tricky practice.

As a challenge for the above obstacles, the Dhahran-Abqaiq road was chosen for a series of studies. It is the first section of the highway running westward from Dhahran and linking the Eastern Province to the major cities in the Gulf States (Fig. 1). This 62-km (38.5 mile) section was upgraded to a four-lane expressway in 1981. The pavement consists of two or more asphalt concrete layers constructed on the subgrade. The total thickness of pavement ranges from 250 to 300 mm (~10 to 12 in.).

[1] Associate professor, Department of Civil Engineering, University of Rhode Island, Kingston, RI 02881.

[2] Director general, Materials and Research Department, Ministry of Communications, Riyadh, Saudi Arabia.

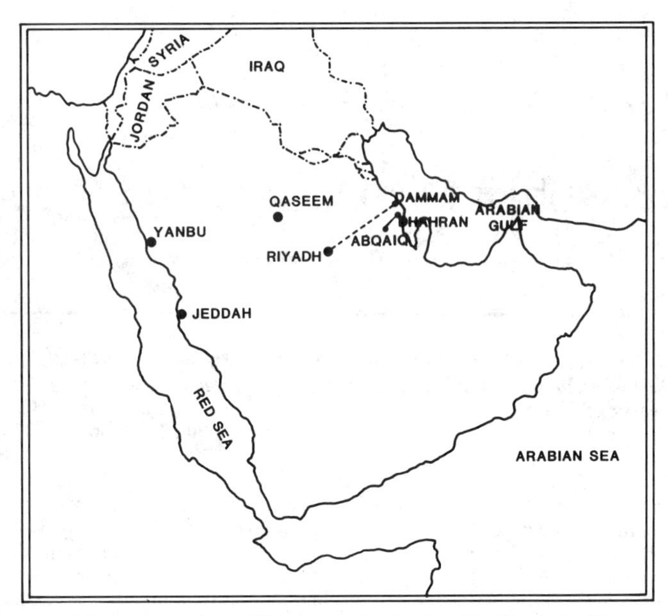

FIG. 1—*Location of Dhahran-Abqaiq Road in Saudi Arabia.*

The source of heavy traffic includes the Damman port, the Dammam Industrial Complex, ARAMCO, Abu-Hadriyyah Rock quarries, and the cement and steel factories in the Eastern Province. The detailed results of the statistical analysis of the gross loads, axle loads, and tire pressures are given in Table 1.

Shortly after the wearing course layer was placed, signs of rutting became visible, mostly on the outside lanes. Various investigations were carried out in the field and laboratory by engineers of the Ministry of Communication (MOC), King Saud University (KSU), and other authorities. The main conclusions were

1. The damage is limited to the wearing course; that is, the subgrade and the base course are intact.
2. The major cause of failure was the inadequacy of the specified mix for the loading and environmental conditions, for example, high temperature.
3. The rate of pavement deterioration was very high.

The common recommendation was immediate rehabilitation.

One of the significant facts was that asphalt mixtures laid down had very high Marshall stability values according to construction records, for example, 1000 to 2000 kg (2205 to 4410 lb). It was doubted that the Marshall stability is a proper indicator, especially when crushed stone is used. Therefore, it had been suggested that the mixture be examined by the Hveem stabilometer. MOC engineers took some cores from the problem areas, tested them by the Hveem stabilometer, and found that they are variant and had much lower values than the common standard criteria, that is, 37 [*2*]. Test results are summarized in Table 2. After undergoing the Hveem stability test, the cores were subjected to the Marshall stability test. The test results indicated that all of them had very high Marshall stability values and a wide range (Table 3), which might indicate that quality control was very poor in the field.

TABLE 1—*Results of statistical analysis on truck loads.*

Axle Sequence No.	Repetitions No.	Mean	Standard Deviation	Minimum	Maximum	Range
				GROSS WEIGHT, KG[a]		
	1161	46 129	14 766	10 950	94 800	83 850
				AXLE WEIGHT, KG		
1	1160	6 070	1304	950	18 950	18 000
2	1160	12 601	3129	1700	20 100	18 400
3	1158	12 120	3208	2400	28 250	25 850
4	682	13 459	4085	2100	20 000	79 000
5	563	14 041	3819	1950	20 900	18 950
6	45	12 899	4231	3400	19 700	16 300
Total	4768	11 191	4285	950	28 250	27 300
				TIRE PRESSURE, PSI[b]		
1	1160	105	18	50	170	120
2	1160	124	18	60	170	110
3	1158	124	18	60	170	110
4	682	129	20	60	180	120
5	563	131	21	60	180	120
6	45	128	23	70	160	90
Total	4768	121	21	50	180	130

[a] 1 kg = 2.2 lb.
[b] 1 psi = 6.89 kPa.

TABLE 2—*Hveem stability test results on cores.*

Serial No.	Specimen No.	Stability, S_c	Serial No.	Specimen No.	Stability, S_c
1	64C	26	20	A4b	32
2	D1b	20	21	C3a	19
3	D6a	28	22	D6b	20
4	B2C	32	23	D3b	21
5	44c	33	24	C2a	13
6	D6d	28	25	C1c	22
7	C4c	25	26	36a	25
8	35c	31	27	23c	12
9	C6a	23	28	54b	34
10	C4a	30	29	71a	20
11	C4b	23	30	A6a	22
12	26b	21	31	D5b	30
13	22c	14	32	65b	14
14	B5c	18	33	62b	24
15	32b	14	34	66b	12
16	A1b	29	35	D4b	27
17	72a	15	36	81c	30
18	85c	22	37	74b	32
19	61b	21	38	66a	26

TABLE 3—*Marshall stability test results on cores.*

Serial No.	Specimen No.	Stability, kg	Serial No.	Specimen No.	Stability, kg
1	64c B	1952	20	A4b B	862
2	D1b W	1667	21	C3a W	922
3	D6a W	1194	22	D6b W	1265
4	B2c W	2613	23	D3b W	1678
5	44c B	1678	24	62a W	1225
6	D6d W	2048	25	C1c B	1681
7	C4c B	1539	26	36a W	1210
8	35c B	738	27	23c W	1198
9	C6a W	937	28	54b W	1620
10	C4a W	1164	29	71a W	1025
11	C4b B	1323	30	A6a W	903
12	26b W	1126	31	D5b W	1471
13	22 C	2414	32	65b W	725
14	B5c W	1867	33	62b W	1073
15	32b B	1259	34	66b W	868
16	A1b B	696	35	D4b W	1284
17	72a B	1114	36	81c B	1697
18	85c W	905	37	74b W	985
19	61b B	992	38	66a W	829

Based on the above facts, a study was designed to compare the Marshall and Hveem mix-design methods using Abqaiq material, which is common in this region. Simultaneously, the sulfur-extended-asphalt (SEA) mixture was examined as one of potential solutions for this rutting problem and was used for a further comparative study. Finally, a decision was made to include test sections in the rehabilitation program of the Dhahran-Abqaiq road, in order to investigate a solution to widespread rutting in the country. Following is a description of the comparative study results and the test sections.

Comparative Study on Mix-Design Methods

The materials, testing procedures, and analysis of test results for the comparative study are described below.

Asphalt Cement

The asphalt cement used in this testing program was a 60 to 70 penetration grade produced by ARAMCO in the Ras Tanurah refinery, Saudi Arabia. This asphalt cement is the same as the one used in Dhahran-Abqaiq road. The properties were examined at the KSU laboratory according to ASTM and AASHTO standards and compared with the General Specifications for Road and Bridge Construction (Saudi Spec, 1972). The test results are given in Table 4 with the Saudi Spec. It was noticed that the penetration value was slightly below the specifications, whereas the kinematic viscosity and ductility values satisfied specifications.

Mineral Aggregates

Mineral aggregates used in the laboratory mixtures of this study were obtained from the construction site of Dhahran-Abqaiq road. It consisted of four components: coarse aggregate (26%), fine aggregate (34%), natural sand (34%), and filler (6%), designated B-1, B-2,

TABLE 4—*Properties of asphalt cement used.*

Characteristics	Test Method AASHTO/ASTM	Test Results	Saudi Specifications
Asphalt Type	[a]	Grade 60–70	Grade 60–70
Producer	[a]	ARAMCO	[b]
Penetration, 25°C 100 g, 5 s (0.1 mm)	T49/D5	54	60–70
Viscosity at 135°C: Centistokes	T201/D2170	598	min. 200
Softening Point (°C)	T53/D36	48	[b]
Ductility, 25°C (cm)	T51/D113	113	min. 100
Flash Point (Cleveland Open Cup), (°C)	T48/D92	315	min. 232.2
Specific Gravity at 25°C (g/cm^3)	T228/D76	1.0347	[b]
Tests on Residues from Thin Film Oven Test	T179/D1754	[a]	[a]
Penetration, 25°C, 100 g, 5 s (0.1 mm)	T49/D5	36	min. 52
Viscosity at 135°C centistokes	T201/D2170	881	[b]
Ductility, 25°C (cm)	T51/D113	103	min. 100

[a] Not applicable.
[b] Saudi Specification does not specify.

B-3, and B-4, respectively. The grain size analysis of aggregates were performed (Table 5). Other tests had been performed to examine various properties of mineral aggregate, and some of the results of which are given in Table 6.

Since the construction gradation was not available, several trial-and-error combinations of aggregate were performed until the satisfactory grading was reached. The comparison between final designed gradation and specifications is shown in the gradation chart of which sieve sizes are raised to 0.45 power (Fig. 2).

TABLE 5—*Grain size analysis of mineral aggregate.*[a]

Seive Size	Components of Aggregates			
	B-1	B-2	B-3	B-4
1.9 cm (3/4 in.)	100	100	100	100
1.3 cm (1/2 in.)	50	100	100	100
No. 4	1	37	100	100
No. 10	0	2	100	100
No. 80	0	0	12	96
No. 200	0	0	1	70

[a] The values indicate the percentage passing the specified sieve.

TABLE 6—*Properties of mineral aggregates.*

General Requirements	Test Method ASTM/AASHTO	Test Results
	B-1	
Bulk specific gravity	C127/T85	2.640
Apparent specific gravity	C127/T85	2.686
Absorption	C127/T85	0.65%
Abrasion	C335/T96	26%
	B-2	
Bulk specific gravity	C127/T85	2.598
Apparent specific gravity	C127/T85	2.674
Absorption	C127/T85	1.1%
Abrasion	C335/T96	26%
	B-3	
Bulk specific gravity	C128/T84	2.608
Bulk specific gravity (SSD)	C128/T84	2.623
Apparent specific gravity	C128/T84	2.649
Absorption	C128/T84	0.6%
	B-4	
Specific gravity	D854/T100	2.708

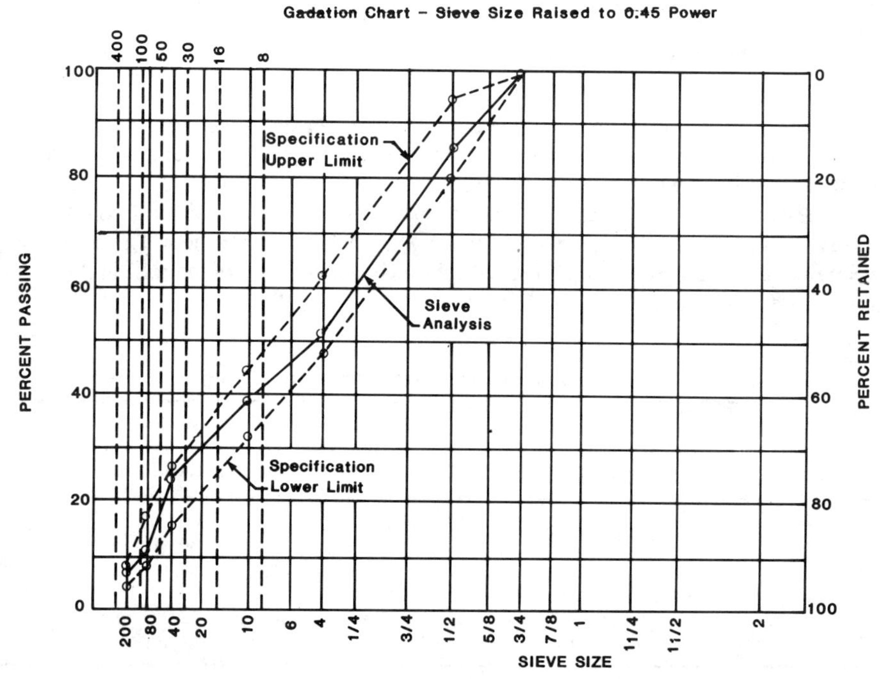

FIG. 2—*Comparison of gradations on chart of which sieve sizes are raised to 0.45 power.*

Asphalt Mixtures

Asphalt mixtures were prepared according to Marshall and Hveem Mix-Design Methods [*2*]. Triplicate Marshall specimens were prepared according to standard procedures for Marshall compaction using 75 blows for 3.5, 4.0, 4.5, 5.0, and 5.5% asphalt content. The specimens removed from the mold were subjected to bulk specific gravity tests after one day, stability and flow tests after two days, and density and void analysis.

Asphalt mixtures for Hveem properties were started with determination of approximate asphalt content by the Centrifuge Kerosine Equivalent (CKE) method. Four Hveem stability test specimens were compacted according to standard procedures for Marshall compaction using 75 blows at each of the following asphalt contents: 3.5, 4.0, 4.5, 5.0. After compaction, specimens contained in the mold were allowed to cool at room temperature for two days. These specimens were used for Hveem stabilometer test and bulk density determination.

Sulfur-Extended-Asphalt (*SEA*)

At standard conditions of temperature and pressure, ordinary sulfur is an odorless and tasteless yellow solid. Its specific gravity is about 2.0 and its melting point is 114°C (238°F) [*3*]. A commercial grade sulfur was used in this study. The SEA composition was established at 30% sulfur and 70% asphalt by weight (18% sulfur and 82% asphalt by volume). This sulfur-to-asphalt ratio was chosen because it is the most common and it replaces a proper amount of the asphalt while still working as a binder that behaved similar to asphalt [*4*].

The KSU procedure for the actual batching of all SEA specimens was a modification of existing procedures [*5–7*]. Since the working range for molten sulfur corresponds quite well to the working range for paving grade asphalt, that is, 124 to 149°C (255 to 300°F) [*7,8*], the sulfur and asphalt were heated separately to about 132°C (270°F). The calculated amounts of the two materials were brought together in a heated stainless steel beaker, and the beaker was placed in a hot oil bath to maintain a constant mixing temperature. The SEA binder was continuously stirred with a propeller until the binder and aggregate were mixed.

SEA Mixtures

The mix-design for SEA mixtures involves substituting an equal volume of SEA binder for the asphalt binder in the conventional mixture. The pre-weighted amount of SEA binder was quickly delivered to hot aggregate in the mixing bowl. The leftover binder was kept in the small can for a penetration test to be done at the same time as the stability test. Mixing was performed the same as for regular asphalt mixtures, and the mixtures were heated to the required compaction temperature for Marshall specimens. Compaction was done according to the standard procedures for Marshall compacting, using 75 blows. After the test specimens were prepared, they were allowed to cool at the room temperature for 2 or 14 days before the Marshall stability and flow tests. The 14-day period was set according to results on maturing study of sulfur-asphalt concrete by Balghunaim [*7*]. He found that the sulfur-asphalt concrete mixes with 30% sulfur required a maturing period of at least 14 days before the stability values start to stabilize. A bulk specific gravity test was performed one day before stability test, and density and void analyses were carried out.

Mixtures for the Hveem specimens were prepared similar to the above Marshall specimens. After compaction, specimens contained in the mold were allowed to cool at room temperature for 2 or 14 days before the Hveem stability test. The bulk density test was performed on the specimens after the completion of the stabilometer tests and as soon as the specimens had cooled to room temperature.

Analysis of Test Results for Asphalt Mixtures

Asphalt mixtures were tested using the Marshall stabilometer for specimens compacted by the Marshall compactor, and the Hveem stabilometer for specimens compacted by Marshall compactor, to compare Marshall and Hveem properties. Specimens used for the Hveem test were also subjected to Marshall tests in order to compare the Hveem results with typical Marshall test results.

Marshall property curves for specimens compacted by a Marshall compactor are shown in Fig. 3. Then the optimum asphalt content was determined from averaging the asphalt contents at maximum stability, maximum unit weight, and 4% air voids for heavy traffic [2]. The optimum asphalt content was found at 4.1%. At this optimum asphalt content, properties of the mixture determined from Fig. 3 were

(*a*) Stability	931 kg (2050 lb)
(*b*) Flow	10
(*c*) Percent air voids	3.5
(*d*) Percent voids in mineral aggregate (VMA)	11.4

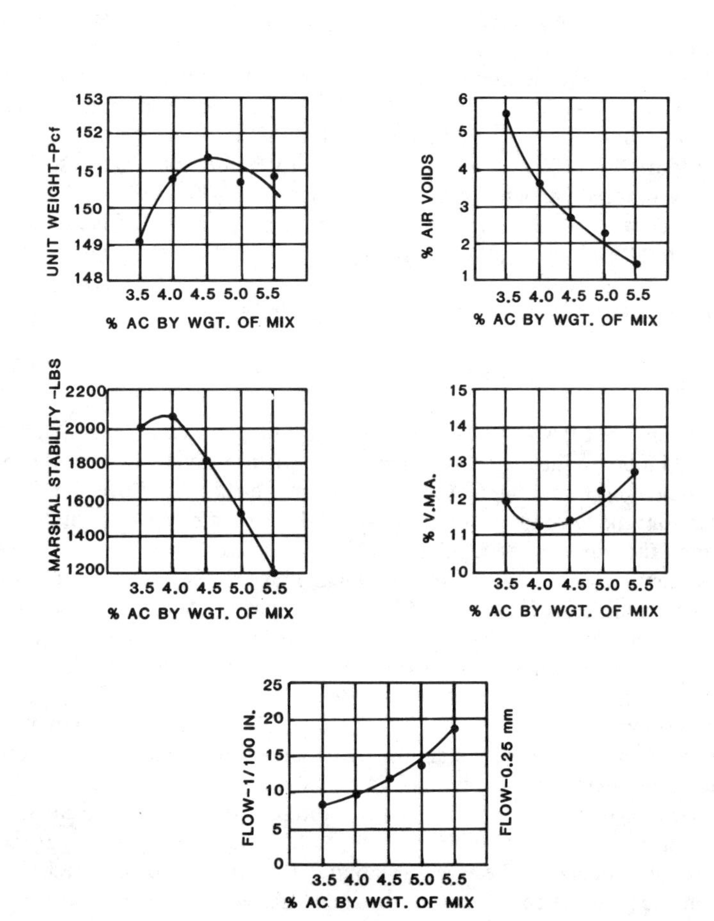

FIG. 3—*Marshall test property curves for asphalt mixtures prepared by Marshall compactor.*

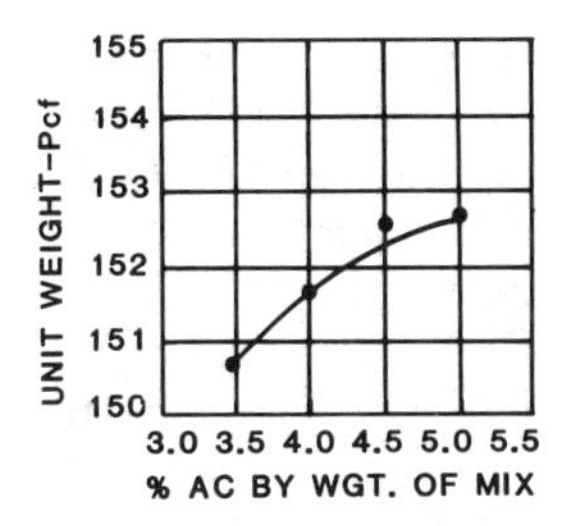

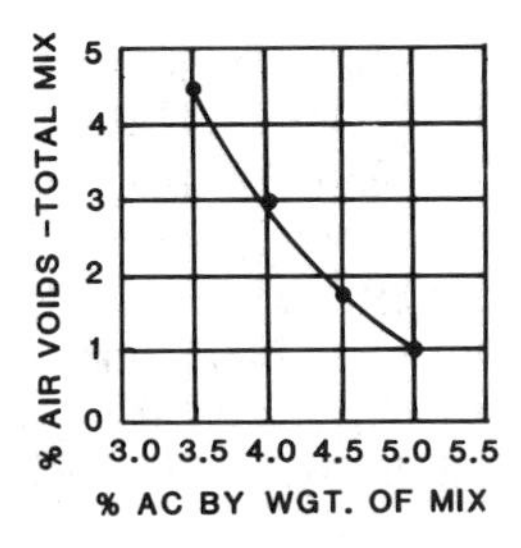

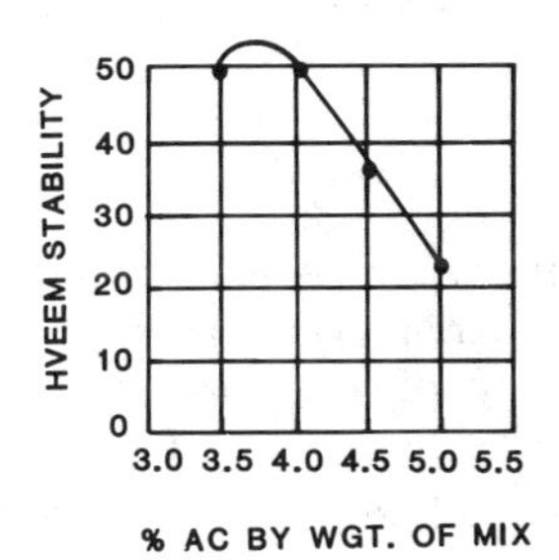

FIG. 4—*Hveem test property curves for asphalt mixtures prepared by Marshall compactor.*

It was noted that the stability value exceeds the minimum of 681 kg (1500 lb), that the flow value is within the limiting range of 8 to 16, and that the percent air voids is within limiting range of 3 to 5. The percent voids in mineral aggregate, however, was less than the minimum of 14%. Since mixture flushing is not expected in the laboratory, the adjustment of gradation to increase VMA values was not attempted. However, a coarser gradation has been suggested and used to increase the VMA values.

Hveem property curves for asphalt mixtures prepared by standard procedures for Marshall compaction in using 75 blows are shown in Fig. 4. The optimum asphalt content was determined from selecting the asphalt contents with no flushing, meeting minimum stability requirements, that is, 37, and 4 or more percent air voids [2]. The optimum asphalt content was 3.5%.

Results (Figs. 3 and 4) indicated that the Marshall and Hveem properties did not correlate and that Hveem mix-design method (3.5%) provided relatively lower optimum asphalt content than Marshall mix-design method (4.1%). It was observed that the Marshall stability value at 3.5% optimum asphalt content determined by the Hveem mix-design method was still high enough to satisfy the Saudi Specifications. Therefore, it may indicate that the Marshall mix-design generally provides the higher optimum asphalt content. It was also observed that specimens of the Hveem mix-design method had higher unit weight and lower percent air voids than ones of the Marshall mix-design method. The higher unit weight and lower air void for specimens of the Hveem test might happen because Hveem test specimens were confined in the mold for two days of curing period and were more or less densified during the Hveem stability test. Further experiments appear to be required to use the Marshall test specimens for Hveem properties. In addition, it can be suggested that Hveem test specimens compacted by Marshall compactor should be extruded like regular Marshall specimen during the period of curing.

It also had been observed from Tables 2 and 3 that all Marshall stability values on the cores satisfied the Saudi Specifications, but all Hveem stability values were less than the Asphalt Institute Criteria [2].

Based on the above observations, it can be concluded that one of the reasons for the rutting at the Abqaiq section was overasphalting (4.61% asphalt content in the field). It also indicates that the Marshall stability test might not be appropriate for the measurement of properties of asphalt mixtures, especially for the crushed limestone. Therefore, it could be a good mix-design practice to consider the Hveem properties for the above similar situations. However, further laboratory performance evaluation is required to verify the above preliminary finding, that is, rutting and fatigue cracking.

Finally, a tighter quality control and coarser gradation can be suggested for the proposed test section and further road construction in Saudi Arabia, based on the study results.

SEA Mixtures

Penetration tests were performed on SEA binder according to the ASTM Test Method for Penetration of Bituminous Materials (D5-83), and it was found that the average value of penetration was 50 after two days of curing and 38 after 14 days of curing. Penetration values against time of SEA are shown in Fig. 5. The large variability of two days' curing SEA might indicate that the maturing process was not completed yet. Specific gravity of SEA cement was estimated as

$$\text{Sulfur specific gravity} = 2.0$$

$$\text{Asphalt specific gravity} = 1.0347$$

$$\text{SEA specific gravity } (G_{\text{SEA}}) = \frac{100}{\dfrac{30}{2.0} + \dfrac{70}{1.0347}} = 1.210$$

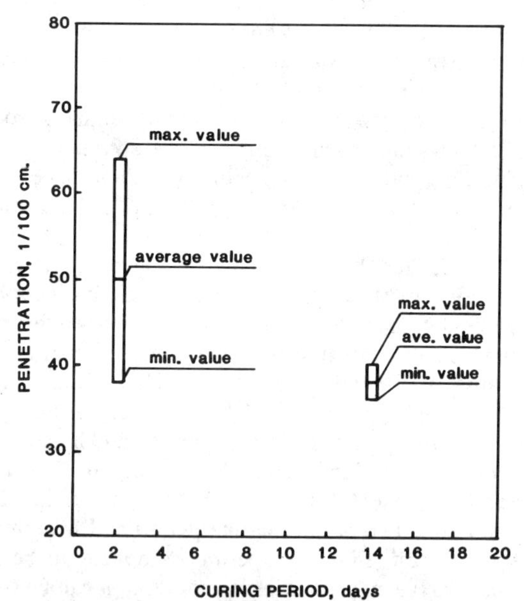

FIG. 5—*Penetration values against time for SEA binder.*

SEA mixtures were tested using Marshall and Hveem stabilometers for specimens prepared by the Marshall compactor. Specimens prepared were cured for 2 and 14 days to compare the two curing periods.

Marshall test property curves for 30/70 SEA mixtures for two days' curing period were prepared as shown in Fig. 6. Then the optimum binder content of 30/70 SEA mixture was determined by the criteria of regular asphalt mixtures. The optimum SEA binder content was found at 4.7% by weight, which has the same volume of the asphalt cement of regular mixture with 4.1% content. At this optimum SEA binder content, properties of the mixture determined from Fig. 6 were

(*a*) Stability	767 kg (1690 lb)
(*b*) Flow	7
(*c*) Percent air voids	3.2
(*d*) Percent VMA	11.3

It was noted that the stability value exceeds the minimum of 681 kg (1500 lb), that the flow value is below the minimum of 8, that the percent air voids is within the limiting range of 3 to 5, and that the VMA was less than the minimum of 14%.

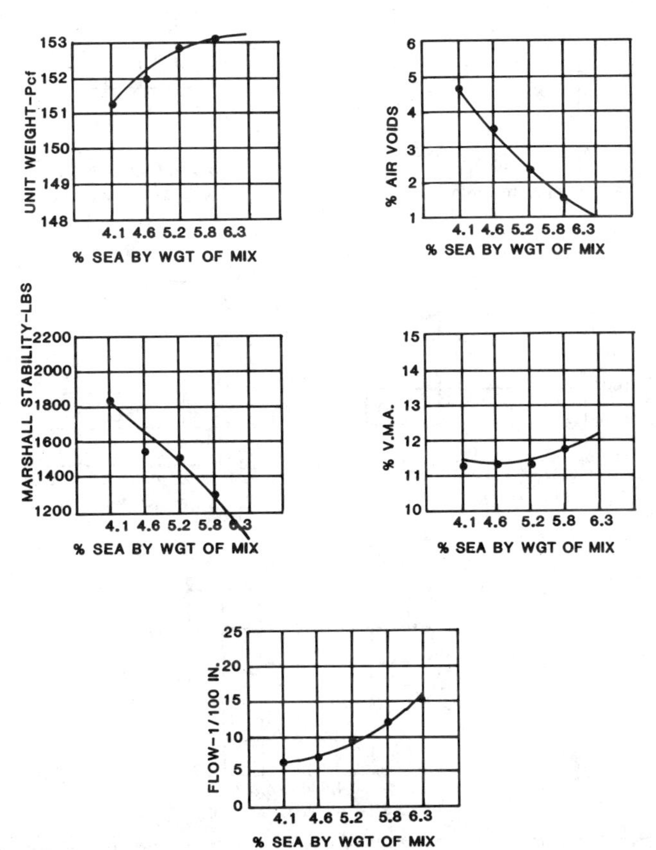

FIG. 6—*Marshall test property curves for 30/70 SEA mixtures prepared by Marshall compactor and cured for two days.*

Marshall test property curves for 30/70 mixture for 14 days curing period were prepared as shown in Fig. 7. The optimum SEA binder content was found at 4.9% by weight, which has the same volume of the asphalt cement of regular mixture with 4.3% content. At this optimum SEA binder content, properties of the mixture determined from Fig. 7 were

(*a*) Stability	999 kg (2200 lb)
(*b*) Flow	10
(*c*) Percent air voids	2.5
(*d*) Percent VMA	11.5

It was noted that the stability value exceeds the minimum of 681 kg (1500 lb), that the flow value is within the limiting range of 8 to 16, that the percent air voids is less than the minimum of 3%, and that the VMA was less than the minimum of 14%.

It was observed that the stability values for SEA specimens cured for 2 days were less than those of specimens cured for 14 days and that other properties were almost the same for both curing periods. Because some of the sulfur in SEA mixture crystallizes with time

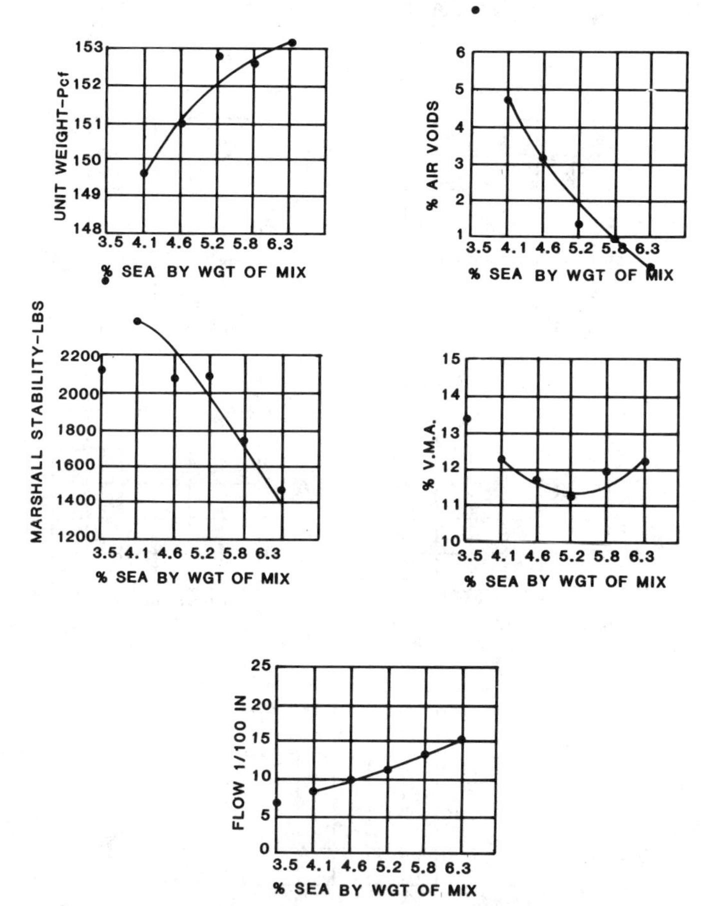

FIG. 7—*Marshall test property curves for 30/70 SEA mixtures prepared by Marshall compactor and cured for 14 days.*

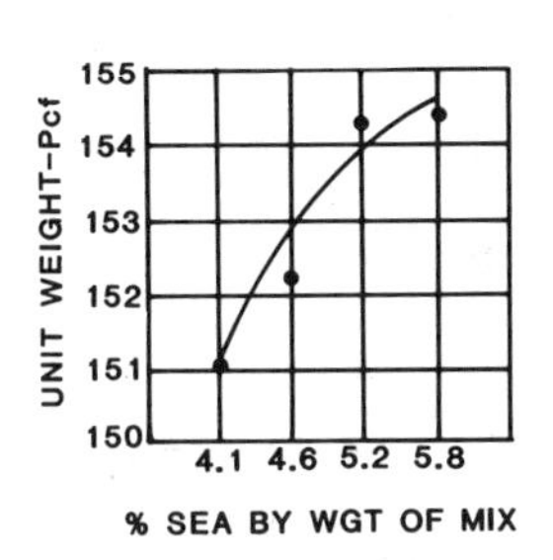

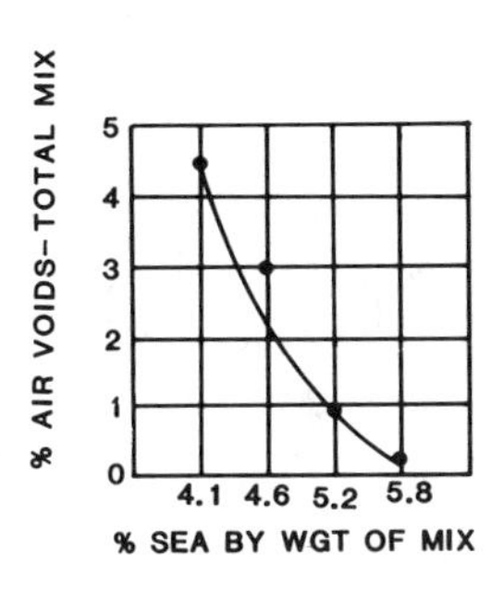

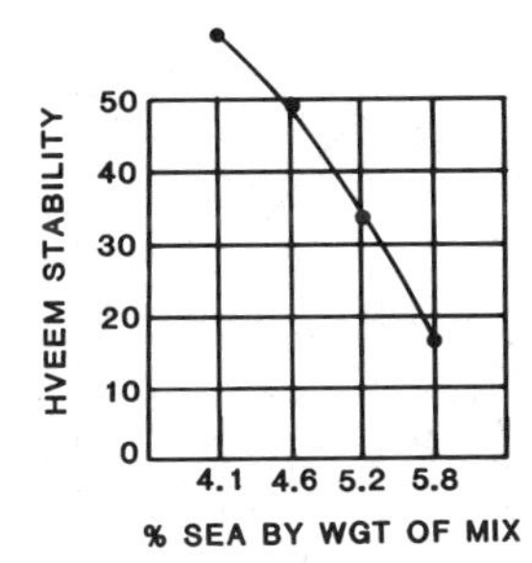

FIG. 8—*Hveem test property curves for 30/70 SEA mixtures prepared by Marshall compactor and cured for two days.*

and acts as a fine material as well as a structuring element, it appears that the increase in stability has occurred. Based on these observations, a decision was made to use properties of SEA mixtures cured for 14 days for further analysis.

Hveem test results and property curves for specimens prepared by Standard procedures for Marshall compactor and cured for two days are shown in Fig. 8. The optimum 30/70 SEA content was determined using the criteria of regular asphalt mixtures [2]. The optimum SEA binder content was found at 4.1% by weight, which has the same volume as the asphalt cement of regular mixtures with 3.5% content.

Hveem test results and property curves for specimens prepared by standard procedures for Marshall compaction using 75 blows and cured for 14 days are shown in Fig. 9. The optimum 30/70 SEA content was found at 4.1% by weight, which has the same volume as the asphalt cement of regular mixtures with 3.5% content.

It was noted that Hveem stability values were not different for SEA mixtures at 2 and 14 days' curing periods. Since both groups were cured in the oven for at least 1 h at 60°C (140°F), that might have caused recrystallization and led to the same Hveem stability values. However, further analysis was based on properties for specimens cured for 14 days as the same as Marshall properties.

Results indicated that Marshall and Hveem properties did not correlate and that the Hveem mix-design method provided relatively lower optimum binder content than the Marshall method. Therefore, the Hveem mix-design method gave higher percent air voids. However, it appears that both mix-design methods can be utilized for 30/70 SEA mixtures with Saudi asphalt and local aggregates in the laboratory. Results also indicated that there

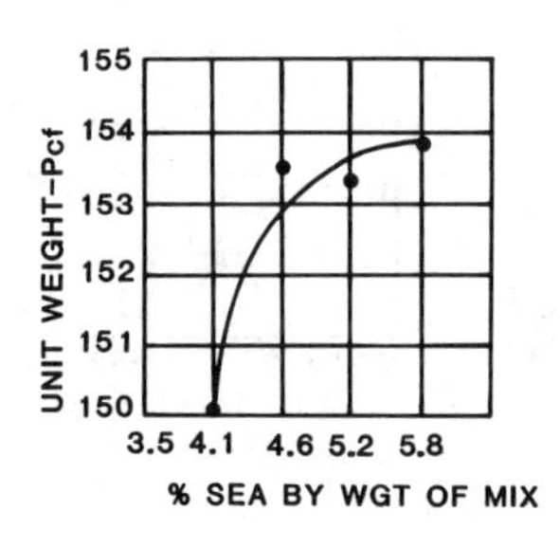

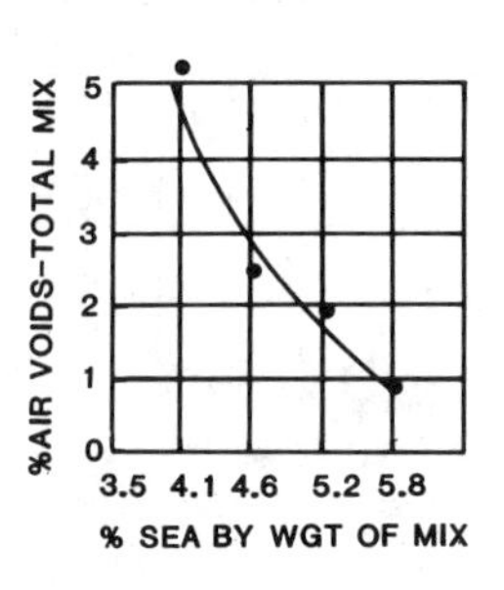

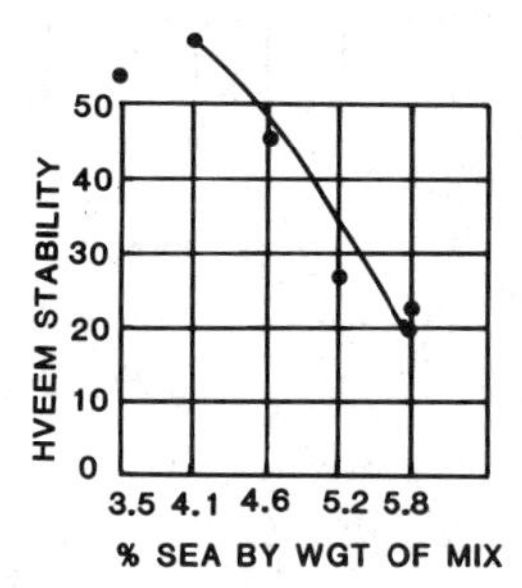

FIG. 9—*Hveem test property curves for 30/70 SEA mixtures prepared by Marshall compactor and cured for 14 days.*

was no significant difference in optimum binder content by volume between asphalt and SEA mixtures for both methods.

For Marshall properties of SEA mixtures it was observed that unit weight values were increased and the maximum was at higher binder content compared with asphalt mix, that voids were decreased, that the Marshall stability was increased and the maximum value was at lower binder content, that there was no significant changes in VMA, and that flow values were decreased at higher SEA binder content. For Hveem properties of SEA mixtures, it was also noted that unit weight values were increased and that there were no significant changes in air voids and Hveem stabilities.

From the comparison between asphalt and SEA mixtures, it was found that SEA mixtures provide higher stability values than asphalt mixtures. Therefore, the utilization of SEA mixtures appears to be feasible where rutting problems are expected.

Rehabilitation Alternatives and Test Sections

Various rehabilitation plans were presented to the decision-makers of MOC. Such plans include

1. Total Rehabilitation

   (*a*) Remove all the wearing course by milling and let traffic drive on the exposed base course.
   (*b*) Lay small sections of trial mixes and evaluate their performance.

(c) At the end of the evaluation period, resurface the entire project with the most suitable mix.

2. Partial Rehabilitation

(a) Mill out the wearing course from the outside lanes where medium to heavy rutting exists, and replace it with four experimental mixtures.
(b) Evaluate mixtures for eight to twelve months under actual traffic.
(c) At the end of the evaluation period, assess the pavement conditions and decide a proper course of action: continuing service or total rehabilitation.

Partial rehabilitation was chosen on the basis of economics, practicality, and the fact that part of the heavy traffic will be diverted through the newly built Riyadh-Dammam Expressway. In addition, the plan will correct the problem and provide valuable information in a very short time.

Test Sections

At least four different wearing course mixes will be placed in previously milled areas and designated as test sections. The length of each section will be 1.5 km (0.9 mile). Based on the behavior of these mixtures during production, placing, and rolling, one will be chosen to replace the wearing course in the remaining milled areas.

The proportions of the mixes are specified in the special specifications. Coarse aggregates are crushed rock from the Abu-Hadriyyah area, separated into four stockpiles of 1.9 cm (3/4 in.), 1.27 cm (½ in.), 0.95 cm (3/8 in.), and 0.64 cm (¼ in.) maximum size. Fine aggregates are crushed stone particles or natural sand passing the 0.95-cm (3/8 in.) sieve. The combined aggregate gradation will be coarser than those specified by the Saudi Specifications.

The mix design will be made using the Marshall apparatus according to Materials and Research Department Method 410. The Marshall Design Criteria with 75 blows are

Marshall stability:	1200 kg (2645 lb) minimum
Marshall flow:	3.5 mm maximum
Air voids:	5% minimum
VMA:	13% minimum
Retained strength:	70% minimum
Asphalt content:	3 to 6%

Specimens prepared at the design asphalt content will be checked by the Hveem stabilometer and for deformation resistance using the Gyratory shear apparatus.

Certain additives, modifiers, and extenders such as sulfur, portland cement, and hydrated lime may be used in portions of the test sections and their effects on mix portions evaluated.

Representative samples will be taken from each day's production. Tests for Marshall stability, bitumen content, aggregate gradations, and voids analysis will be performed and checked against the job mix formula. Corrections will be made when warranted.

The spatial specifications will detail the milling, mixing, placing, and compaction operations and equipment. Approximately 25% of the cost of the project will go to furnishing, installing, and maintaining work area signs, temporary pavement markings, concrete barriers, delineators, and lighting devices necessary to provide for smooth and safe flow of traffic in the work areas. The details are clearly specified and illustrated in the special specifications and the addenda.

Certain specialized and more fundamental tests will be carried out on fabricated specimens during construction and on pavement cores during the evaluation period. The structural integrity of the pavement will be periodically monitored using deflection measuring devices. Continuous traffic counts will be made. Axle and gross weight measurements will be made to characterize the traffic loading.

The information and expertise gained from this project will help the MOC improve its specifications, improve its construction and quality control methods, and provide adequate solutions to the widespread rutting problem. In addition, this project will introduce a new methodology in problem-solving, that is, practical research.

Summary and Conclusions

Based on the findings from this study, the following conclusions are drawn.

1. Hveem mix-design was successfully performed on specimens prepared by standard procedures for Marshall compaction.
2. Test results indicated that the Marshall and Hveem properties did not correlate and that Hveem mix-design method provided relatively lower optimum binder content than Marshall Method for this particular mixture.
3. Some of the potential reasons for the rutting at Dhahran-Abqaiq road were overasphalting, fine-graded mix, and poor quality control.
4. The Marshall Mix-Design method for asphalt mixture was successfully used for 30/70 SEA mixtures in this study. The Hveem Mix-Design method for specimens compacted with Marshall compactor was also successfully utilized in finding Hveem properties of SEA mixtures with Saudi asphalt and local aggregates in the laboratory.
5. It was observed that the stability values for SEA specimens cured for 2 days were less than those of specimens cured for 14 days and that other properties were almost the same for both curing periods. The average value of penetration was 50 after 2 days and 38 after 14 days of curing time.
6. Test results indicated that there was no significant difference in optimum content by volume between asphalt and SEA mixtures for both methods.
7. It was found that 30/70 SEA mixtures had higher densities and lower air voids than pure asphalt mixtures for Marshall method. It was also noted that SEA mixtures had higher unit weights and that there was no significant change in air voids for the Hveem method.
8. It was found that SEA mixtures give higher stability values than asphalt mixtures. Therefore, the utilization of SEA mixtures appears to be feasible where rutting problems are expected.
9. To provide a solution to widespread rutting in Saudi Arabia, a decision was made to include test sections in the rehabilitation program of the Dhahran-Abqaiq road. Trial mixture will be prepared with relatively less binder content and coarser gradation than the problem mixture with and without additives.

Acknowledgments

The authors wish to thank the transportation material group at King Saud University and the engineers of the Saudi Ministry of Communication for their help in performing various tests and in formulating rehabilitation alternatives. Special appreciation is extended to Mr. Fares M. Al-Sarhani for his assistance in conducting laboratory experiments and to Professors Thomas W. Kennedy and William D. Kovacs for their suggestions in preparing this manuscript. We also would like to express our thanks to the Department of Civil Engineering, University of Rhode Island, which provided the typing and drafting support required for this paper.

References

[1] Pearson-Kirk, D., "Techniques to Reduce Road Accidents Involving Trucks," *Proceedings*, Safety Techniques and Applications Symposium, King Saud University, Riyadh, Saudi Arabia, 1985.
[2] "Mix-Design Methods for Asphalt Concrete and Other Hot-Mix Type," Manual Series No. 2, The Asphalt Institute, 1979.
[3] Izatt, J. O. and Gallaway, B. M., "Sulphur-Extended-Asphalt Field Trials, MH 153 Brazos County, Texas," FHWA Interim Report No. TX-78-536-2, Federal Highway Administration, Dec. 1978.
[4] McCullagh, F. R., "Using a Dry-Drum in the Construction of Sulphur-Extended-Asphalt (SEA) Pavements," FHWA-TS-80-243, Federal Highway Administration, Aug. 1980.
[5] Kennedy, T. W. and Hass, R., "Sulphur-Asphalt Pavement Technology: A Review of Progress," Transportation Research Record 741, Transportation Research Board, 1980, pp. 42–49.
[6] Gallaway, B. M. and Epps, J. A., "Updating Technology Aspects of Sulphur and Its Usage in Highway Paving," *Proceedings*, 1st Arab Regional Conference on Sulphur and Its Uses in the Arab World, April 1982.
[7] Balghunaim, F. A. R., "Evaluation of Maturing and Moisture Susceptibility of Sulphur-Asphalt Concrete," Ph.D. Dissertation, The University of Michigan, Ann Arbor, MI, 1984.
[8] McBee, W. C., Sullivan, T. A., and Izah, J. O., "State of the Art Guideline Manual for Design, Quality Control, and Construction of Sulphur-Extended-Asphalt (SEA) Pavements," FHWA-IP-80-14, Federal Highway Administration, Aug. 1980.

James R. Lundy,[1] R. Gary Hicks,[1] and Robert McHattie[2]

Evaluation of Percent Fracture and Gradation on the Behavior of Asphalt Concrete Mixtures

REFERENCE: Lundy, J. R., Hicks, R. G., and McHattie, R., **"Evaluation of Percent Fracture and Gradation on the Behavior of Asphalt Concrete Mixtures,"** *Implication of Aggregates in the Design, Construction, and Performance of Flexible Pavements, ASTM STP 1016,* H. G. Schreuders and C. R. Marek, Eds., American Society for Testing and Materials, Philadelphia, 1989, pp. 120–143.

ABSTRACT: The effects of percent fracture and fines content on the laboratory performance of asphalt mixtures were investigated. The objective of this study was to develop an approach for the effective utilization of crushed aggregates in Alaskan highway and airfield pavements. In addition, current Alaskan specifications were evaluated in light of this laboratory testing.

The repeated load diametral test device was used to measure the mixture performance in terms of modulus, permanent deformation, and fatigue. Experimental variables included percent fracture, percent passing the 200 sieve, and aggregate source. Testing was conducted at temperatures representative of Alaskan environmental conditions.

Test results show optimum asphalt contents to be minimized at approximately 8%. At the temperatures tested, modulus changes could not be attributable to fracture levels. Significant reductions in laboratory lives were noted when fines contents were varied from the 6% level currently specified.

Laboratory testing indicates the current specifications for fines content is appropriate and, in fact, the mid-range value of 6% maximizes fatigue lives when standard Alaskan pavement sections were analyzed. At the temperatures tested, little increase in fatigue life can be attributed to increase in fracture.

KEY WORDS: bituminous concrete, flexible pavements, aggregates, fine content, crushed faces, fatigue life, modulus

At the time of writing (1986), crushed aggregates are specified differently in Alaska depending on whether they are intended for highway or airfield applications.[3] Except for the minus No. 200 (fines) content in highway base courses, crushed aggregate specifications are not tied directly to Alaska's present pavement design methods; therefore, there is no way to permit design variabilities for the actual load-bearing capabilities of nonspecification materials. This paper presents an evaluation of the ability to relate aggregate properties to their expected performance in flexible pavement sections. The objectives of this paper are:

[1] Instructor and professor, respectively, Department of Civil Engineering, Oregon State University, Corvallis, OR 97331.

[2] Regional materials engineer, Alaska Department of Transportation and Public Facilities, 2301 Peger Road, Fairbanks, AK 99701.

[3] "Standard Specifications for Highway Construction" (1981) and "Standard Specifications for Airfield Construction" (1982), both Alaska Department of Transportation and Public Facilities.

1. To develop a soundly based approach for the utilization of crushed aggregates in highway and airfield pavements, and
2. To provide a rational link between highway and airfield pavement designs and their utilization of crushed aggregates.

Alaskan aggregates were tested at three fracture and gradation combinations in asphalt concrete mixtures. The effects of these variables on mixture performance were compared using various repeated load tests.

Test Program

The purpose of the laboratory study was to determine the effect of percent fracture and gradation (minus No. 200) on the properties of asphalt concrete mixtures. This section of the paper describes the variables studied, materials used, material processing performed, and the specimen preparation and test procedures used.

The asphalt concrete mixtures were tested using dynamic test procedures to determine properties which could serve as inputs for multilayered pavement analysis computer programs. In this way, the effect of material properties on pavement performance could also be studied. Therefore, the tests performed were designed to represent field conditions as closely as possible.

Variables Considered

The test variables evaluated in the laboratory are presented in Table 1. Their selection was based on Alaska Department of Transportation and Public Facilities (ADOTPF) specifications and on a review of the literature [1]. Three aggregate sources (Anchorage, Fairbanks, and Juneau) were selected for study. These sources were chosen so that the study would represent materials which are used throughout Alaska. The current ADOTPF specifications for aggregates used in aggregate base and asphalt concrete require a minimum fracture level of 70% as determined by Alaska Test Method T-4.[4] In order to bracket the specification over a reasonable range, test conditions of 50, 70, and 90% fracture were used.

The aggregate gradations used are given in Table 2. For all materials larger than the No. 40 screen, the mean of the ADOTPF gradation limits was used (Fig. 1). The minus No. 200 quantities studied are the high limit, low limit, and mean value from the specifications. Note that only the gradation of the material passing the No. 40 screen was varied in this study.

Aggregates—The Anchorage aggregates were obtained from the Anchorage Sand and Gravel, Palmer Terminal Pit. The Fairbanks aggregates came from the Sealand Pit stockpile.

TABLE 1—*Variables and levels of treatment considered in laboratory experiments.*

Variable	Level of Treatment
Aggregate source	Anchorage, Fairbanks, Juneau
Level of fracture, %	50, 70, 90
Gradation (−200 sieve), %	3, 6, 10

[4] All Alaska Test Methods are from *Alaska Test Methods,* Materials Section, State of Alaska, Department of Transportation and Public Facilities, Jan. 1980.

TABLE 2—*Aggregate gradation.*

Sieve Size	Percent Passing[a]
¾ in.[b]	100
⅜ in.	80
No. 4	55
No. 8	40
No. 40	20
No. 200	3, 6, 10

[a] Mean grading for Alaska Type II asphalt concrete.
[b] 1 in. = 2.54 cm.

The Juneau aggregates were from the Juneau Ready Mix, Upper Lemon Creek stockpile. These materials were pit-run river gravels.

Table 3 presents the aggregate specific gravity and durability data which were received from the material sources. The State of Alaska specifies that aggregates used in asphalt concrete mixtures shall conform to the durability requirements contained in the American Association of State Highway and Transportation Officials (AASHTO) standard material specifications M-283 [*2*]. These requirements classify aggregates into three quality classes based on the percent wear as determined by Los Angeles Abrasion Machine. Class A has a maximum percent wear of 40%, Class B, 45% maximum, and Class C, 50% maximum. For all three classes, the maximum percent loss as determined by sodium sulfate soundness testing is 12%. Table 3 shows that all three aggregates are classified as Class A aggregates. Since the performances of all three aggregates in this testing were greatly in excess of the durability requirements, it appears reasonable to assume that degradation of the aggregates would not be a factor in this laboratory test program.

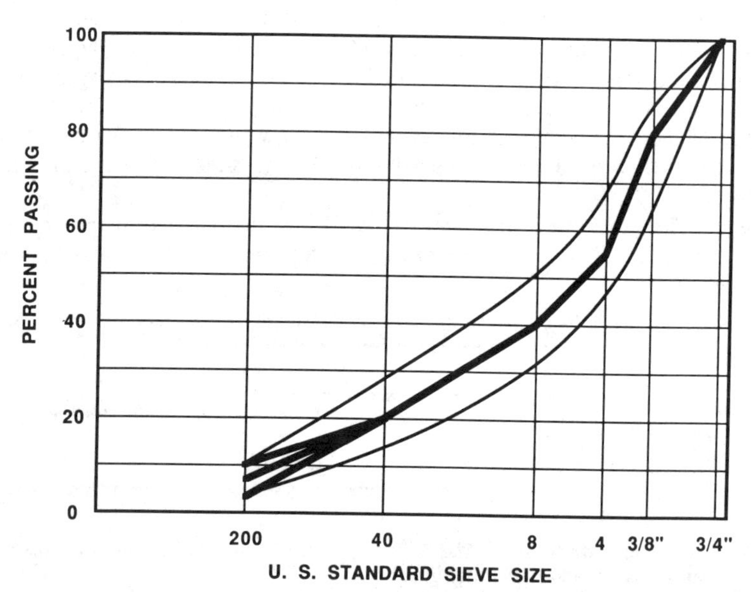

FIG. 1—*Alaska Type II gradation specification band and gradations used.*

TABLE 3—*Aggregate properties.*

Materials	Specific Gravity Apparent	Bulk	SSD	Sodium Sulfate Soundness	Los Angeles Abrasion Percent Wear, % Loss	Grade	Degradation, %
Anchorage fine	2.77	2.54	2.66	1	...	...	...
Anchorage coarse	2.80	2.77	2.78	1	15	A	68
Fairbanks fine	...	...	...	3	...	...	...
Fairbanks coarse	2.70	2.64	2.66	1	29	A	87
Juneau fine	2.78	...	...	3	...	...	...
Juneau coarse	2.75	...	...	1	33	A	77
Alaska specifications				12 max	40 max		45 min

Asphalt—The asphalt used was an AC-5 grade supplied by Chevron USA. The asphalt properties provided by Chevron are given in Table 4, along with the appropriate specification limits.

Material Processing and Preparation

The aggregates sent from all three sources differed significantly in fracture level and gradation. This necessitated processing each aggregate separately in order to establish uniformity among the three. A description of the materials received and processing performed on each is presented in the following paragraph.

The goal of the processing was to separate out and store a quantity of aggregate in each particle-size range and at each fracture level for each source. In this way, the materials could be recombined to the necessary gradation and fracture level for each test. All materials were first oven-dried and then separated in a large-sieve series. The sizes used were 2.54, 1.91, 0.95 cm (1.0, 0.75, 0.37 in.), Nos. 4, 8, 40, and 100 sieves. Material passing the No. 40 but retained on the 100 and material passing the No. 100 sieve were sampled and wet-sieved over a No. 200 screen. The two materials were then blended to obtain the desired passing 200 contents. Representative specimens of material larger than the No. 4 sieve were evaluated for fracture level using Alaska Test Method T-4. Percent fracture was determined by the ratio of fractured aggregates to nonfractured particles. An aggregate is determined to be fractured when one face is broken.

The Anchorage source sent two graded aggregates. One was a crushed gravel with a fracture level of 90% and the other an uncrushed gravel with a fracture level of 10%. These were blended together to achieve the desired fracture levels. The Fairbanks source sent two

TABLE 4—*Asphalt properties for AC-5.*

Property[a]	Actual	Specification
Viscosity, 140°F,[b] poises	458	500 ± 100
Viscosity, 275°F,[b] cSt	161	110 min
Penetration, 77°F,[b] dmm	147	120 min
Flashpoint, COC, °F[b]	595	350 min

[a] Initial asphalt properties.
[b] °C = (°F − 32) × 5/9.

graded aggregates which had fracture levels of 50 and 70%. To achieve the 90% fracture level, it was necessary to crush some of the material in a laboratory jaw crusher. The Juneau source sent a graded aggregate with a 70% fracture level and a clean gravel with negligible fracture. The three fracture levels were obtained by a combination of crushing and blending.

Specimen Preparation

Laboratory specimens were prepared using the Marshall Procedure as described in Alaska Test Method T-17. Aggregates and molds were heated in a thermostatically controlled oven for at least 8 h. Heated asphalt was added to the aggregate to produce the desired binder content. Mixing took place at 135°C (275°F) for 2 min, followed by a 50 blow-per-side compaction when the temperature had reached 120°C (248°F). After cooling to room temperature, the specimens were extracted. All samples were prepared as described previously.

Tests and Procedures

Optimum asphalt contents were determined for each of the variable combinations considered using the Marshall mix design method [*2*]. Optimum content was determined based on the criteria shown in Table 5. Marshall flow and stability testing was accomplished using an MTS testing machine. An *X-Y* plotter was used to record maximum load and deformation during the test. From these plots, flow and stability were computed.

Bulk specific gravity was determined using AASHTO T-165. Voids filled and percent air voids were calculated based on the theoretical maximum specific gravity as determined from AASHTO Method T-209.

All modulus, fatigue, and split tension tests were performed at 10°C (50°F). This temperature is most representative of average conditions in Alaska. Twelve specimens were prepared for each variable combination at an asphalt content 0.4% above optimum. A common practice in Alaska is to increase the asphalt content used in mixtures by 0.4% to insure adequate film thicknesses. This practice was followed on all variable combinations provided the criterion limits (see Table 5) were not exceeded. Nine specimens were used in modulus and fatigue testing, with three reserved for split tension testing.

Resilient modulus tests were performed using the apparatus shown in Fig. 2. This device allows the monitoring of recoverable horizontal deformation using two horizontally mounted linear variable differential transducers (LVDT). In addition, applied loads are monitored with a load cell. Equation 1 was used to calculate the resilient modulus [*2*]

TABLE 5—*Criteria for selection of optimum asphalt contents.*[a]

Parameter	Allowable Range
Maximum unit weight	...
Maximum stability	1000 lb/min[b]
Voids filled of 80%	70 to 90
Voids total mix of 3%	2 to 5
Flow value, 1/100 in.[c]	8 to 16

[a] Alaska Test Method T-17.
[b] 1 in. = 2.54 cm.
[c] 1 lb = 4.448 N.

FIG. 2—*Test equipment used to measure modulus of asphalt concrete specimens.* (a) *Overview of system with environmental cabinet.* (b) *Closeup of diametral yoke with specimen.*

$$M_R = \frac{P}{\Delta H \times t}(0.2692 + 0.9974\nu) \tag{1}$$

where

M_R = resilient modulus, psi,
P = dynamic load, lb,
t = thickness of specimen, in.,
ΔH = total horizontal instantaneous recoverable deformation, in., and
ν = Poisson's ratio.

Poisson's ratio was assumed constant and equal to 0.35, allowing Eq 1 to be simplified to

$$M_R = \frac{0.6183 \times P}{\Delta H \times t} \tag{2}$$

Three specimens were tested at each of three microstrain levels: 50, 75, and 100. These values were chosen to insure failure in a reasonable length of time and allow fatigue-strain relationships to be developed.

After moduli were determined, the specimens were subjected to constant stress repeated loading until failure. For this study, failure was defined by using a thin aluminum strip placed around the horizontal specimen diameter perpendicular to the direction of loading (Fig. 3). As the specimen deforms, this strip is gradually strained until it breaks, stopping the test. Permanent vertical deformations were recorded as the fatigue test progressed using dial gage indicators.

In addition to the tests described previously, static indirect tension tests were performed on three specimens for each variable combination considered. Again, the MTS device was used with the crosshead speed set at 5.08 cm/min (2 in./min). These tests were also conducted at 10°C (50°F). Tensile strength was calculated using [2]

$$S_t = \frac{2(P_{max})}{\pi t d} \tag{3}$$

where

S_t = tensile strength, psi,
P_{max} = load at failure, lb,
t = thickness of specimen, in., and
d = diameter of specimen, in.

Results for Surfacing Materials

Mix Design

The mix designs for each combination of variables considered are summarized in Table 6. Values of design asphalt content were plotted against percent passing the No. 200 sieve at constant fracture levels and against percent fracture at a constant fines content for Anchorage aggregates in Fig. 4. Similar trends were associated with Fairbanks and Juneau aggregates.

FIG. 3—*Test setup for fatigue and deformation tests.* (a) *Fatigue specimen.* (b) *Instrumentation used to measure permanent deformation.*

TABLE 6—*Optimum asphalt contents.*[a]

Aggregate Source	Percent Fracture	Percent Passing 200 sieve	Percent Asphalt (by weight of aggregate)
Fairbanks	50	6	5.5
	70	3	6.2
		6	5.4
		10	5.7
	90	3	6.2
		6	5.9
		10	6.2
Anchorage	50	6	5.2
	70	3	6.4
		6	4.9
		10	4.7
	90	3	6.3
		6	5.6
		10	5.1
Juneau	50	6	5.5
	70	3	6.8
		6	5.7
		10	5.6
	90	3	7.4
		6	6.1
		10	6.3

[a] Reflects Alaskan 0.4% increase over standard optimum.

Modulus and Fatigue Results

The results of modulus and fatigue testing are summarized in Tables 7 through 9 for Fairbanks, Anchorage, and Juneau, respectively. The modulus and fatigue life values represent the average of three specimens at each strain level. Previous work [*3–5*] has established that the number of repetitions to failure can be related to tensile strain level using

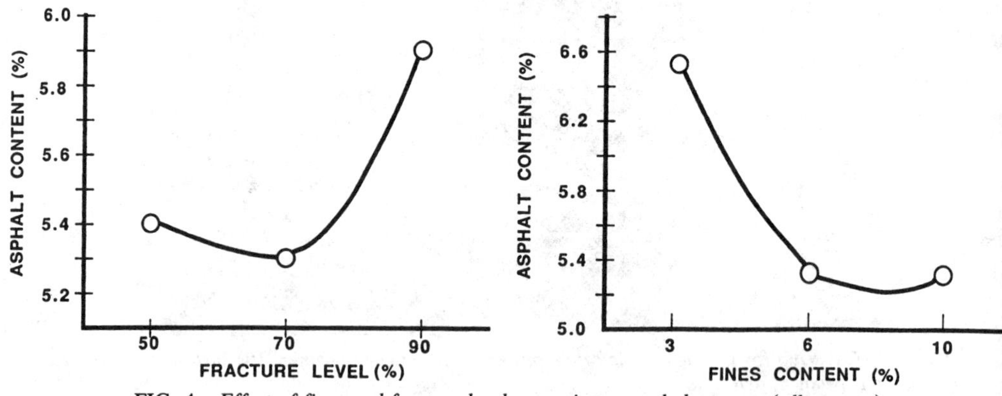

FIG. 4—*Effect of fines and fracture level on optimum asphalt content (all sources).*

TABLE 7—*Modulus and fatigue results—Fairbanks.*

Percent Fracture	Percent P200	Microstrain	Average Modulus, ksi[a]	Average Fatigue Life
50	6	100	1518	3638
		75	1628	4949
		50	1639	15 040
70	3	100	1242	3249
		75	1390	4260
		50	1438	13 964
	6	100	1607	2956
		75	1713	5868
		50	1874	23 455
	10	100	1092	10 902
		75	1321	15 508
		50	1400	81 001
90	3	100	1283	3530
		75	1286	5383
		50	1425	11 933
	6	100	1608	3086
		75	1568	9570
		50	1806	20 580
	10	100	1371	3710
		75	1730	4082
		50	1481	30 475

[a] 1 ksi = 6.895 kPa.

TABLE 8—*Modulus and fatigue results—Anchorage.*

Percent Fracture	Percent P200	Average Microstrain	Average Modulus, ksi[a]	Fatigue Life
50	6	100	1471	2731
		75	1561	4368
		50	1754	10 088
70	3	100	1141	3591
		75	1204	6782
		50	1171	20 392
	6	100	1564	2784
		75	1529	7892
		50	1743	19 192
	10	100	1502	3312
		75	1541	8978
		50	1595	33 350
90	3	100	1345	3720
		75	1210	8525
		50	1441	17 981
	6	100	1523	3564
		75	1580	9217
		50	1834	22 865
	10	100	1836	3309
		75	1835	8402
		50	1884	27 950

[a] 1 ksi = 6.895 kPa.

TABLE 9—*Modulus and fatigue results—Juneau.*

Percent Fracture	Percent P200	Microstrain	Average Modulus, ksi[a]	Average Fatigue Life
50	6	100	1395	2662
		75	1467	4561
		50	1461	10 803
70	3	100	1151	2242
		75	1170	5148
		50	1227	11 997
	6	100	1368	2940
		75	1251	5024
		50	1313	15 422
	10	100	1249	3917
		75	1333	9034
		50	1450	23 937
90	3	100	979	3942
		75	955	8272
		50	1049	16 961
	6	100	1159	3634
		75	1224	6569
		50	1330	24 215
	10	100	1024	5070
		75	1049	7709
		50	1233	18 264

[a] 1 ksi = 6.895 kPa.

the power curve relationship

$$N_f = k(\epsilon_t)^c \tag{4}$$

where

N_f = repetitions to failure,
ϵ_t = initial tensile strain, in./in., and
k, c = regression coefficients.

Least-squares regression analysis was performed on each variable combination data set to determine k, c, and the coefficient of determination, R^2. The results of the regression analysis are presented in Table 10.

In addition, Fig. 5 was included to allow the direct comparison of results for Juneau aggregates. The effect of fracture level and fines contents on fatigue life was similar for all aggregate sources tested.

Permanent Deformation

Permanent deformation was recorded during fatigue testing for each sample. To allow meaningful comparisons between various data sets, permanent strain (ϵ_p) was plotted against number of repetitions (N) at the initial stress levels for each variable combination. Initial stress was calculated using Kennedy's equation (Eq 3). The applied load was used in place of P_{max} to determine the initial stress.

TABLE 10—*Summary of regression data.*

Aggregate Source	Percent Fracture	Percent P200	Regression Coefficients k	c	R^2
Fairbanks	50	6	2.88 E^{-5}	−2.01	0.84
	70	3	6.42 E^{-6}	−2.16	0.83
		6	2.47 E^{-9}	−3.01	0.98
		10	2.32 E^{-8}	−2.90	0.89
	90	3	3.58 E^{-4}	−1.75	0.92
		6	1.10 E^{-7}	−2.62	0.89
		10	9.43 E^{-10}	−3.12	0.86
Anchorage	50	6	9.99 E^{-5}	−1.86	0.92
	70	3	2.67 E^{-7}	−2.53	0.97
		6	9.05 E^{-8}	−2.63	0.92
		10	1.72 E^{-10}	−3.32	0.89
	90	3	4.81 E^{-6}	−2.23	0.91
		6	2.22 E^{-7}	−2.56	0.90
		10	3.03 E^{-9}	−3.01	0.95
Juneau	50	6	3.47 E^{-5}	−1.97	0.89
	70	3	2.51 E^{-5}	−2.02	0.92
		6	6.98 E^{-7}	−2.40	0.94
		10	5.19 E^{-7}	−2.47	0.89
	90	3	2.75 E^{-5}	−2.04	0.92
		6	8.07 E^{-8}	−2.66	0.92
		10	3.22 E^{-4}	−1.80	0.88

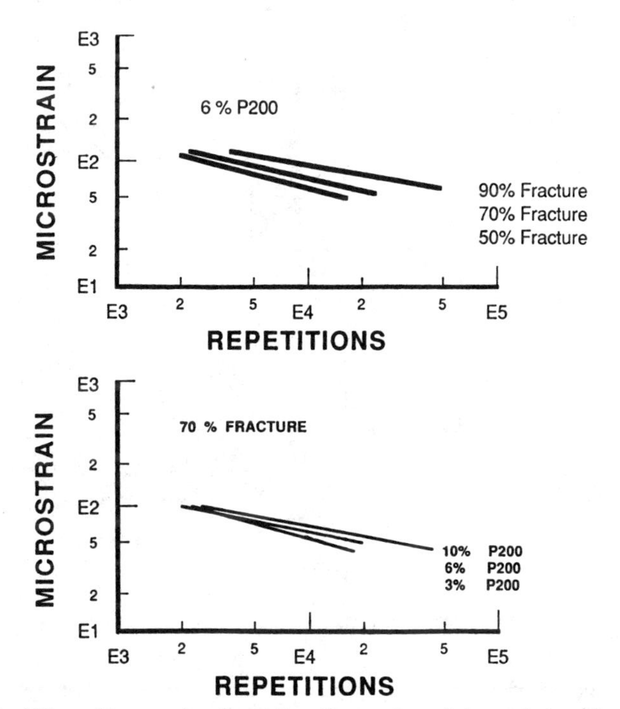

FIG. 5—*Effect of fracture* (top) *and fines* (bottom) *on fatigue relationship—Juneau.*

The plot of repetitions versus permanent deformation showing the effect of fines content is included in Fig. 6 for Juneau. A similar plot for effect of fracture is shown in Fig. 7. Using these data and their companion least-squares regression curve fits, it was possible to estimate the expected permanent deformation at each of three initial stress levels for a given number of loadings. Plots were then constructed of initial stress versus permanent deformation at 10^3 and 10^4 repetitions. Example data for 10 000 repetitions showing the effect of fines content and fracture level are included in Figs. 8 and 9, respectively. These figures allow comparison of the effect of different fracture and P200 levels after a similar number of loadings.

Split Tension

The results of split tension testing are included in Table 11 for each of the aggregate sources tested. As noted previously, tests were conducted at 10°C (50°F) using a crosshead rate of 5.08 cm/min (2 in./min). The tensile strength in Table 11 was computed according to Eq 3. To demonstrate the effect of fracture and percent fines on split tensile strength, Fig. 10 was developed.

Evaluation of Results

Mix Design

Examination of the optimum asphalt contents presented in Fig. 4 shows the influence of gradation and fracture to be complex. With a constant P200, the asphalt contents necessary

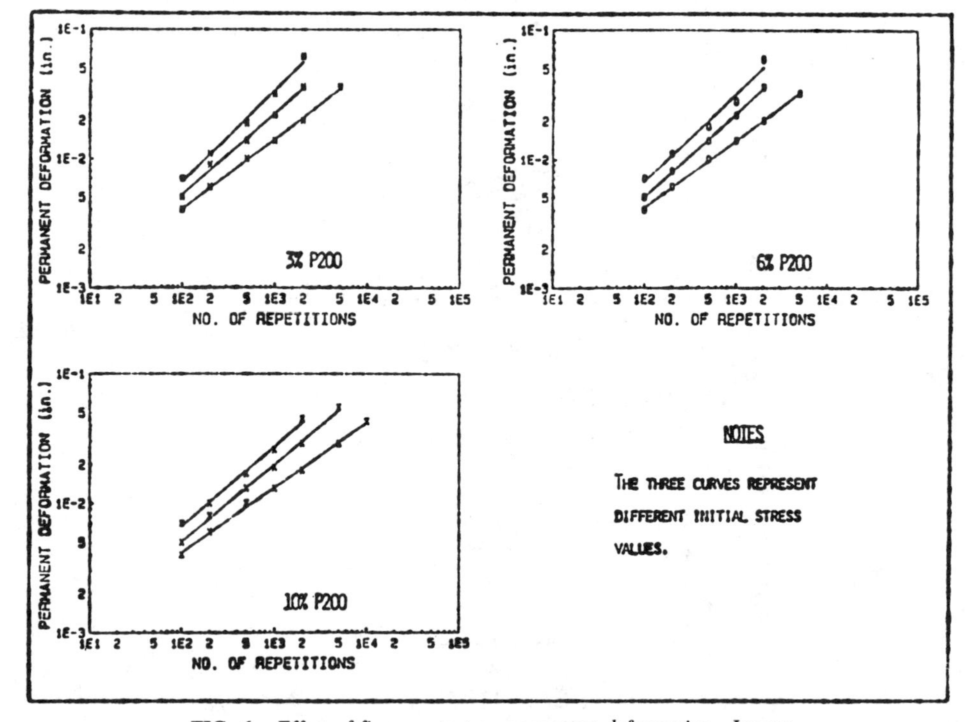

FIG. 6—*Effect of fines content on permanent deformation—Juneau.*

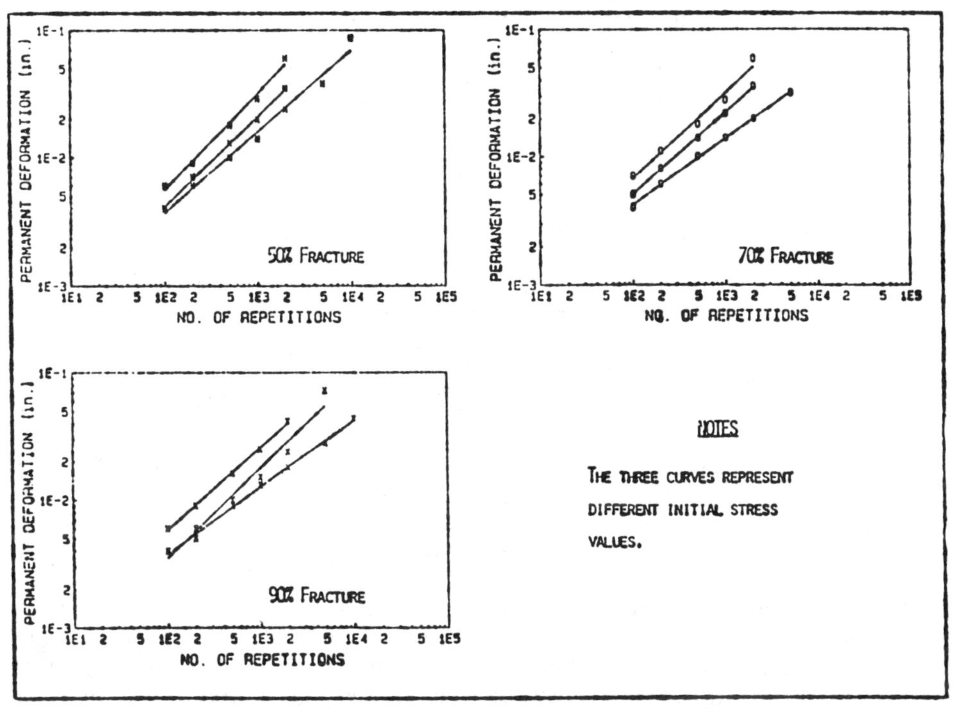

FIG. 7—*Effect of fracture on permanent deformation—Juneau.*

to meet Alaskan mix design criteria decreased slightly between 50 and 70% fracture level, then increased with an increasing percent fracture. This trend may reflect the increased roughness of the fractured aggregates when compared to the uncrushed particles.

The effect of increasing fines content varied with increasing P200 content. The optimum asphalt content decreases as the fines content changes from 3 to 6%. The decrease in asphalt necessary to achieve optimum content is a result of the criteria used, in particular, total voids. Voids may be decreased by adding asphalt cement or by increasing the fines content. As the fines content is increased from 3 to 6% the total voids decreased without requiring

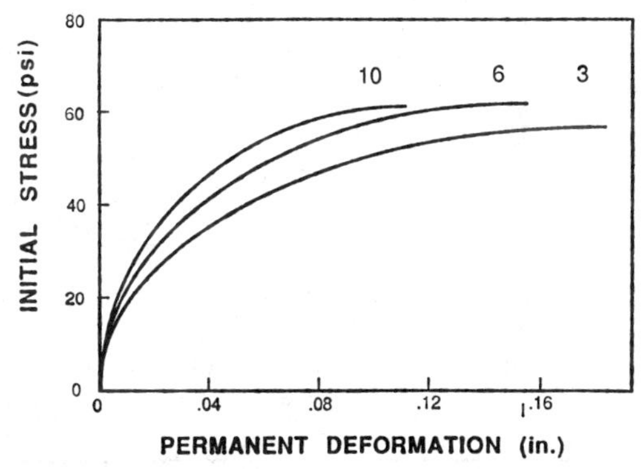

FIG. 8—*Variation of permanent deformation with initial stress—effect of fines—Juneau.*

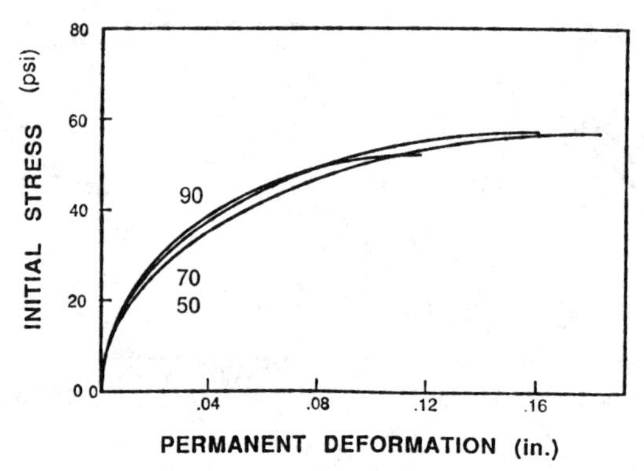

FIG. 9—*Variation of permanent deformation with initial stress—effect of fracture—Juneau.*

additional asphalt. In fact, the required asphalt decreased. However, as the fines are increased from 6 to 10%, the trend reverses and additional binder is needed.

Resilient Modulus

Modulus data have been summarized in Tables 7 through 9 for each of the aggregate sources included in this study. Examination of these data shows few consistent trends in modulus variability with either gradation or fracture change. The differences in optimum asphalt contents among sample sets have apparently masked the influence of fracture and gradation.

TABLE 11—*Split tension test results.*

Source	Percent Fracture	Percent P200	Tensile Strength, psi[a]
Fairbanks	50	6	326
	70	3	278
	70	6	306
	70	10	265
	90	3	261
	90	6	308
	90	10	272
Anchorage	50	6	288
	70	3	302
	70	6	325
	70	10	327
	90	3	302
	90	6	346
	90	10	371
Juneau	50	6	335
	70	3	275
	70	6	340
	70	10	330
	90	3	274
	90	6	334
	90	10	302

[a] 1 psi = 6.895 Pa.

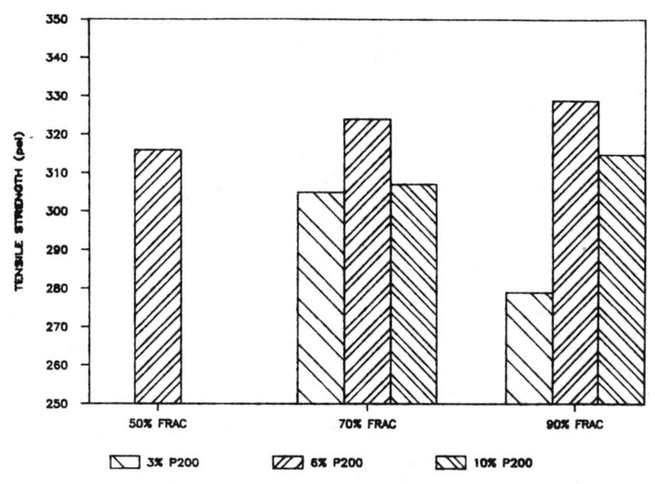

FIG. 10—*Average split tensile strength—all sources.*

The Fairbanks data allow limited comparative evaluation of the effects of gradation and fracture at common binder contents. Optimum asphalt contents at the 90% fracture level for both 3 and 10% fines are the same. The increase in fines results in a 15% increase in resilient modulus. This increase in modulus is attributed to the interaction of asphalt and P200 material to produce a more viscous binder [*6–10*].

Examination of the modulus values presented show the moduli to be higher than normally expected for standard Alaska mixes tested at 10°C (50°F). In an attempt to explain these discrepancies, additional tests were performed on specimens after fatigue testing to determine the asphalt content, gradation, and recovered binder properties. The results of these tests are presented in Table 12. These values were used to estimate the expected modulus values utilizing Eq 6, developed by the Asphalt Institute.

TABLE 12—*Recovered asphalt properties and gradation of Juneau—70% fracture specimens.*

Specimen ID: Percent Passing Sieve Size	J3	J6	J10
¾ in.[a]	100	100	100
⅜ in.[a]	87	81	80
No. 4	65	58	57
No. 10	40	40	40
No. 40	22	22	22
No. 200	4.2	7.5	11.3
Extracted asphalt content, %	6.6	6.4	5.6
viscosity, kinematic 275°F,[b] Centistoke	217	219	220
Viscosity, absolute 140°F,[b] poises	760	800	811
Penetration at 77°F[b] (cm/100)	78	98	91

[a] 1 in. = 2.54 cm.
[b] °C = (°F − 32) × 5/9.

$$\log |E^*| = 5.553833 + 0.028829 \left(\frac{P_{200}}{f^{0.17033}}\right) - \text{-}0.03476(V_v)$$

$$+ 0.070377 (\eta_{70°F}) + 0.000005 [t_p(1.3 + 0.49825 \log f)\ _{\text{Pac}}0.5]$$

$$- 0.00189 \left[t_p{}^{(1.3+0.49825 \log f)} \frac{\text{Pac}^{0.5}}{f^{1.1}}\right] + 0.931757 \left(\frac{1}{f^{0.02774}}\right) \quad (6)$$

where

$|E^*|$ = stiffness of asphalt concrete, psi,
P_{200} = percent aggregate passing No. 200 sieve,
f = load frequency, H_z,
$\eta_{70°F}$ = absolute viscosity at 70°F, poises × 10^6,
V_v = percent air voids,
Pac = asphalt content, percent by total weight of mix, and
t_p = temperature, °F.

This regression equation has an average relative error of 22.9% between predicted and laboratory measured modulus values [7].

For this study, the frequency of loading was maintained at 1 Hz, allowing simplification of Eq 6 to

$$\log |E^*| = 6.485590 + 0.028829(P_{200}) - 0.001885[t_p{}^{1.3}\ \text{Pac}^{0.5}] \quad (7)$$

The results of the calculations using Eq 7 are presented in Table 13 along with the laboratory measured results. The comparison shows reasonable agreement between measured and calculated values.

It should be noted that the absolute viscosity at 21.1°C (70°F) was not measured from the recovered asphalt properties. This value was estimated using Figs. 11 and 12 in conjunction with recovered penetration and viscosity values.

TABLE 13—*Comparison of measured and predicted modulus values—Juneau.*

Percent Fracture	Percent P200	Modulus, ksi[a]		
		Measured	Predicted[b]	Percent Difference
50	6	1441	1361	−6
70	3	1183	901	−24
	6	1376	1191	−13
	10	1344	1803	34
90	3	994	957	−4
	6	1238	1327	7
	10	1102	1746	58
			Average percent difference = 7%	

[a] 1 ksi = 6.895 kPa.
[b] Predicted values computed using recovered asphalt properties and Eq 7.

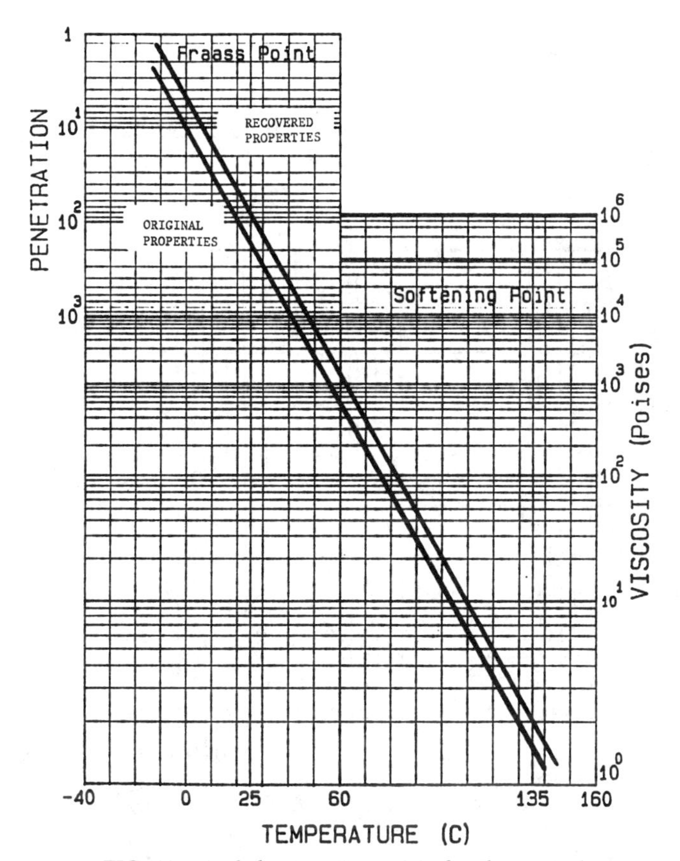

FIG. 11—*Asphalt properties—original and recovered.*

Fatigue Tests

The influence of fracture level on fatigue life is presented graphically in Fig. 5. This plot indicates that for the range of fracture level investigated in this study, laboratory fatigue lives were increased approximately 75% by increasing fracture levels from 50 to 90%. However, this benefit appears to be present only at lower strain levels (<150 μSt). All testing was performed at 10°C (50°F). At higher temperatures, reduced binder viscosities would allow for more aggregate displacements. These displacements tend to enhance the aggregate interlock feature of higher fracture levels [*2*].

Figure 5 demonstrates the effect of percent passing the No. 200 sieve on the fatigue lives of laboratory specimens. Data from all aggregate sources show few consistent trends in the influence of fines content on fatigue life. The effect of binder content may be masking the importance of fines content, particularly at the lowest P200 level. Generally, the 3% P200 mixes contained an average of 1% additional binder.

Permanent Deformation

The effect of percentage passing the No. 200 sieve on permanent deformation is displayed in Fig. 8. This shows a reduction in permanent deformations with increasing fines at 10 000

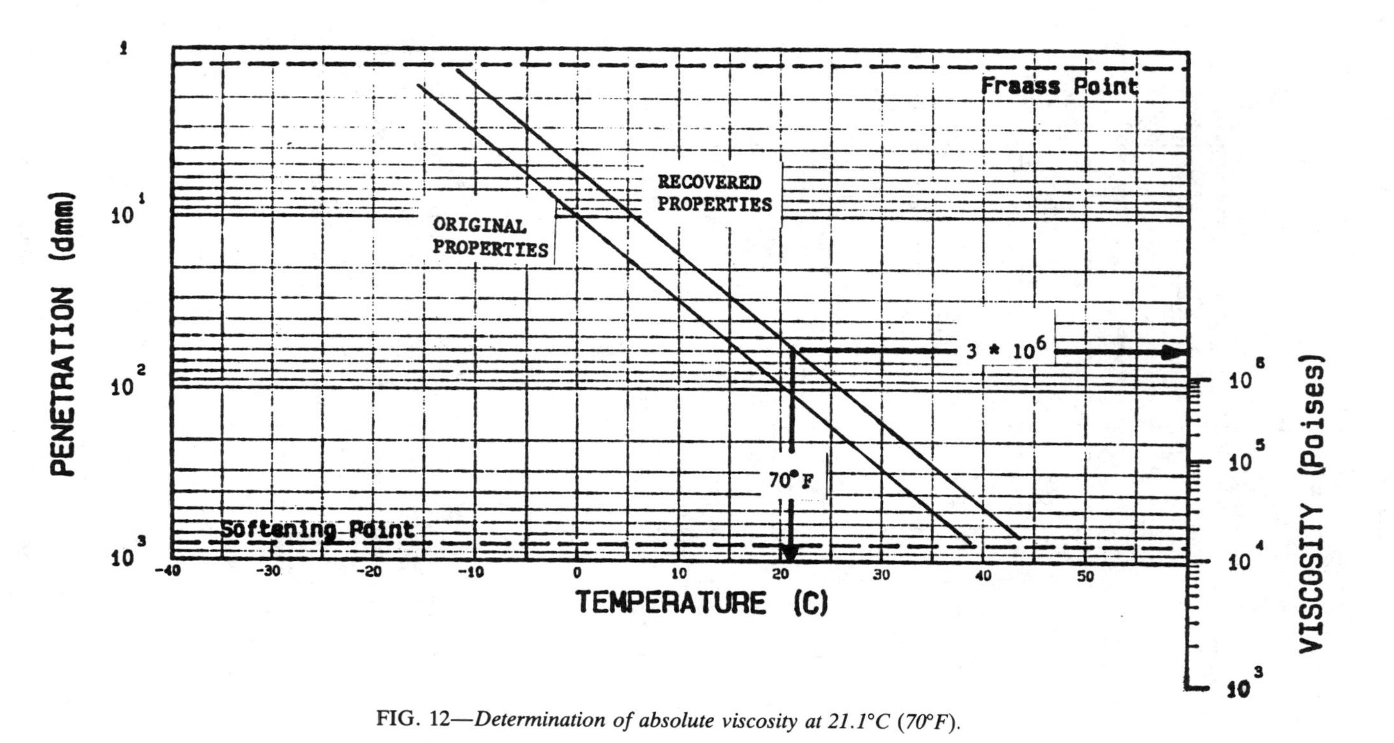

FIG. 12—*Determination of absolute viscosity at 21.1°C (70°F).*

repetitions. Generally, as the asphalt content increases, the permanent deformation associated with a given initial stress level also increases.

The effect of fracture on permanent deformation values is presented in Fig. 9. The results for Juneau show little reduction in permanent deformation with increasing fracture at a given initial stress level when tested at 10°C (50°F). The results, however, are not consistent across the aggregate types tested. Anchorage aggregates show the greatest reduction in permanent deformations with increasing fracture. These results may be attributed to the lower asphalt contents associated with all Anchorage variable combinations. Optimum asphalt contents of Anchorage aggregates averaged 0.6% lower than those from Juneau or Fairbanks. The lower binder contents allows the development of interparticle interlock of the increasingly fractured aggregate which more effectively resists deformation.

Split Tension

The results of split tension testing (Fig. 10) show few recognizable trends regarding the effect of fracture level, aggregate type, or percent fines. Specimens of 6% fines display a slight increase in tensile strength with an associated increase in fracture level. As noted previously, all laboratory testing was conducted at 10°C (50°F). These low temperatures, combined with the different binder contents, would likely mask the effect of increasing fracture levels.

Summary of Results

The effects of fracture and fines content on optimum asphalt content as determined by this study show that an increase in required binder results from increasing fracture levels. An increase in fines content from 3 to 6% reduces the required asphalt cement; however, if fines content were to increase from 6 to 10%, the required binder would increase slightly. This may be attributed to the total voids criterion used in mix design. Voids can be reduced by increasing fines from 3 to 6% without increasing binder; however, additional P200 material necessitates increased asphalt cement.

Differences in optimum asphalt contents among the variable combinations considered may effectively mask the function of fracture level and fines content in both resilient modulus and fatigue life. From the data obtained in this study, little benefit may be ascribed to increasing the fracture level in order to prolong fatigue life at high strain levels (>75 μSt); however, at higher temperatures the effect of fracture may be greater.

As with fatigue and modulus, permanent deformation results indicate limited differences between the variable combinations considered. Some reduction in permanent deformation resulted from increasing the fines content from 3 to 6%. However, this increase is accompanied by a decrease in binder content, which may account for the reduction.

Split tensile strength results show the influence of increasing fracture at both the 6 and 10% fines content. These large strain tests activate the aggregate interlock mechanisms in the specimens. Increased interlock can be associated with the increased fracture. This aggregate interlock effect would also be activated by testing at increased temperatures.

Effect of Fracture and Fines Content on Pavement Performance

Typical Alaskan pavement cross sections were modeled using the Elastic Layered System computer program (ELSYM5). This program determines the stresses, strains, and displacements in a three-dimensional ideal elastic layered system. A more detailed description of this program and its limitations is provided in Ref *11*.

Input variables required by ELSYM5 are

1. Loadings, including wheel load, tire pressure, and location of loads.
2. Modulus, Poisson's ratio, and thickness for each layer.

Cross sections and loadings used to analyze the surfacing materials are described in Table 14. Assumed conditions include the following:

1. Base moduli were constant for each of the seasonal conditions examined.
2. Surface moduli varied with fines content.
3. Surface moduli varied with change in temperature from spring to summer, as shown in Table 14.

Little difference in modulus was attributed to fracture level variation from 50 to 90% at the temperature tested. Fines content affected the modulus and this variation was incorporated in the surfacing moduli used in the ELSYM5 analysis.

TABLE 14—*ELSYM5 inputs for analysis.*

Material	Thickness, in.[b]	Poisson's Ratio	Percent P200	Modulus Values, ksi[a]		
				Spring No. 1	Spring No. 2	Summer
			HIGHWAY SECTION			
AC	2.4	0.35	3	1450	1450	500
			6	1800	1800	600
			10	1600	1600	550
Base	4.8	0.35	...	12.5	12.5	25.0
Borrow	36	0.35	...	5.0	5.0	18.0
Subgrade	24	0.35	...	50.0	1.5	12.5
	∞	0.35	...	50.0	50.0	12.5
Loading: two uniform circular loads of 5000 lb each at three radii center-to-center; 80-psi contact pressure.						
			AIRFIELD SECTION			
AC	4	0.35	3	1450	1450	500
			6	1800	1800	600
			10	1600	1600	550
Base	16	0.35	...	12.5	12.5	25.0
Borrow	48	0.35	...	5.0	5.0	18.0
Subgrade	24	0.35	...	50.0	1.5	12.5
	∞	0.35	...	50.0	50.0	12.5
Loading: two uniform circular loads of 76 900 lb each at four radii center to center; 160-psi contact pressure.						

[a] 1 ksi = 6.895 kPa.
[b] 1 in. = 2.54 cm.
[c] 1 lb = 0.45 kg.

Results

Results of the ELSYM5 analysis are presented in Table 15 along with the pavement cross sections considered. The strains shown represent the maximum tensile strain at the bottom of the surfacing layer. These tensile strains were input into an appropriate fatigue life equation to yield the expected fatigue life of the surfacing.

TABLE 15—*Laboratory fatigue lives based on ELSYM5 analysis.*[a]

	Thickness, in.[b]				Modulus, ksi[c]					Maximum Tensile Strain at Bottom of Surfacing, E-6 in./in.	Percent Change in Fatigue Life
Loading	Surface	Base	Borrow	Spring Thaw	Surface	Base	Borrow	Thaw	Subgrade		
Highway summer	2	4	36	...	500	25	18	...	12.5	326.8	−66
	2	8	36	...	500	25	18	...	12.5	307.3	−66
	4	4	36	...	500	25	18	...	12.5	267.4	−72
	4	8	36	...	500	25	18	...	12.5	257.0	−72
	2	4	36	...	600	25	18	...	12.5	315.3	...
	2	8	36	...	600	25	18	...	12.5	297.0	...
	4	4	36	...	600	25	18	...	12.5	245.3	...
	4	8	36	...	600	25	18	...	12.5	235.8	...
	2	4	36	...	550	25	18	...	12.5	321.1	−52
	2	8	36	...	550	25	18	...	12.5	302.2	−50
	4	4	36	...	550	25	18	...	12.5	255.6	−49
	4	8	36	...	550	25	18	...	12.5	245.8	−47
Highway spring No. 1	2	4	36	...	1450	12.5	5	...	50	414.4	−66
	2	8	36	...	1450	12.5	5	...	50	370.7	−70
	4	4	36	...	1450	12.5	5	...	50	211.1	−77
	4	8	36	...	1450	12.5	5	...	50	201.3	−77
	2	4	36	...	1800	12.5	5	...	50	388.6	...
	2	8	36	...	1800	12.5	5	...	50	332.6	...
	4	4	36	...	1800	12.5	5	...	50	180.7	...
	4	8	36	...	1800	12.5	5	...	50	173.2	...
	2	4	36	...	1600	12.5	5	...	50	393.3	−56
	2	8	36	...	1600	12.5	5	...	50	353.1	−59
	4	4	36	...	1600	12.5	5	...	50	196.8	−47
	4	8	36	...	1600	12.5	5	...	50	188.0	−45
Highway spring No. 2	2	4	36	24	1450	12.5	5	1.5	50	420.7	−70
	2	8	36	24	1450	12.5	5	1.5	50	374.3	−70
	4	4	36	24	1450	12.5	5	1.5	50	220.0	−77
	4	8	36	24	1450	12.5	5	1.5	50	207.2	−77
	2	4	36	24	1800	12.5	5	1.5	50	375.5	...
	2	8	36	24	1800	12.5	5	1.5	50	336.6	...
	4	4	36	24	1800	12.5	5	1.5	50	188.9	...
	4	8	36	24	1800	12.5	5	1.5	50	178.8	...
	2	4	36	24	1600	12.5	5	1.5	50	399.8	−62
	2	8	36	24	1600	12.5	5	1.5	50	356.9	−59
	4	4	36	24	1600	12.5	5	1.5	50	205.3	−48
	4	8	36	24	1600	12.5	5	1.5	50	193.8	−46
Airfield	4	16	48	...	1450	12.5	5	...	50	570.3	−62
	4	16	48	...	1800	12.5	5	...	50	537.0	...
	4	16	48	...	1600	12.5	5	...	50	555.6	−66
	4	16	48	24	1450	12.5	5	1.5	50	596.2	−61
	4	16	48	24	1800	12.5	5	1.5	50	564.1	...
	4	16	48	24	1600	12.5	5	1.5	50	582.1	−67
	4	16	48	...	500	25	18	...	12.5	228.0	−59
	4	16	48	...	600	25	18	...	12.5	248.3	...
	4	16	48	...	550	25	18	...	12.5	235.9	−33

[a] Uses the average regression coefficients from the 3%, 6%, and 10% P200, respectively. Base case was assumed to be the 6% P200 situation.

[b] 1 in. = 2.54 cm.

[c] 1 ksi = 6.895 kPa.

Fatigue life equations used are shown below for each of the fines contents tested:

P200	Fatigue Equation
3	$7.03\ E^{-5} \times \epsilon_t^{-2.122}$
6	$1.83\ E^{-5} \times \epsilon_t^{-2.413}$
10	$1.09\ E^{-7} \times \epsilon_t^{-2.964}$

To allow more meaningful comparison of the results, the 6% P200 was assumed to be the reference case since that represents the current specification. Percent differences from that base condition were calculated and are given in Table 15.

Conclusions and Recommendations

The following conclusions summarize the important findings of this study:

1. When tested at 10°C (50°F), fatigue lives did not appear to be significantly affected by fracture level between 50 and 90% at high strains; however, fatigue lives were significantly affected at lower strain levels.
2. Optimum asphalt contents as determined by the Alaskan Marshall mix design require minimum binder contents when fines content is between 7 and 9%.
3. Modulus tests show no consistent changes with fracture level; however, maximum moduli occur at approximately 6% fines.
4. Reduction in permanent deformation associated with increasing fracture levels was not demonstrated when tested at 10°C (50°F).
5. When tested at 10°C (50°F), split tension values show no effect due to fracture or fines content.

The following recommendations concerning changes to the current design and specification procedures for the utilization of crushed aggregates are based on the results of this study:

1. The current specifications for fracture and fines content in asphalt concrete appear appropriate for highways. Changes are not recommended.
2. Current airfield fracture specifications seem to be unnecessarily high based on laboratory testing.
3. Additional tests should be conducted at other temperatures to quantify the effects of fracture level and fines content on asphalt concrete mixtures.
4. Future testing should attempt to quantify the effect of asphalt content on modulus and fatigue results.

References

[1] Albright, S. A., Lundy, J. R., and Hicks, R. G., "Evaluation of Percent Fracture and Gradation on Behavior of Asphalt Concrete and Aggregate Base—Final Report," Transportation Research Report 85-17, Transportation Research Institute, Oregon State University, Dec. 1985.

[2] Kennedy, T. W., "Characterization of Asphalt Pavement Materials Using the Indirect Tensile Test," *Proceedings*, Association of Asphalt Paving Technologists, Vol. 46, 1977, pp. 132–150.

[3] Pell, P. S., "Characterization of Fatigue Behavior," Special Report 140, Highway Research Board, 1973, pp. 49–65.

[4] Witczak, M. W., "Fatigue Subsystem Solution for Asphalt Concrete Airfield Pavements," Special Report 140, Highway Research Board, 1973, pp. 112–130.

[5] Havens, J. H., Deen, R. C., and Southgate, H. F., "Pavement Design Scheme," Special Report 140, Highway Research Board, 1973, pp. 130–142.

[6] Pell, P. S. and Taylor, I. F., "Asphaltic Road Materials in Fatigue" in *Proceedings,* Association of Asphalt Paving Technologists, Vol. 38, 1969, pp. 371–442.

[7] Epps, J. A. and Monismith, C. L., "Influence of Mixture Variables on the Flexural Fatigue Properties of Asphalt Concrete," *Proceedings,* Association of Asphalt Paving Technologists, Vol. 38, 1969, pp. 423–464.

[8] Pell, P. S. and Cooper, K. E., "The Effect of Testing and Mix Variables on the Fatigue Performance of Bituminous Materials," *Proceedings,* Association of Asphalt Paving Technologists, Vol. 44, 1975, pp. 1–37.

[9] Shook, J. F. and Kallas, B. F., "Factors Influencing Dynamic Modulus of Asphalt Concrete," *Proceedings,* Association of Asphalt Paving Technologists, Vol. 38, 1969.

[10] Dukatz, F. and Anderson, D. A., "The Effect of Various Fillers on the Mechanical Behavior of Asphalt Concrete," *Proceedings,* Association of Asphalt Paving Technologists, Vol. 49, 1980, pp. 530–549.

[11] Hicks, R. G. and McHattie, R. L., "Use of Layered Theory in the Design and Evaluation of Pavement Systems," Report No. FHWA-AK-RD-83-8, Alaska Department of Transportation and Public Facilities, July 1982.

Ali A. Selim[1] *and Najim A. Heidari*[2]

Measuring the Susceptibility of Emulsion Based Seal Coats to Debonding

REFERENCE: Selim, A. A. and Heidari, N. A., **"Measuring the Susceptibility of Emulsion Based Seal Coats to Debonding,"** *Implication of Aggregates in the Design, Construction, and Performance of Flexible Pavements, ASTM STP 1016,* H. G. Schreuders and C. R. Marek, Eds., American Society for Testing and Materials, Philadelphia, 1989, pp. 144–153.

ABSTRACT: Asphalt mixes, hot and cold, can experience bond loss due to moisture damage, a phenomenon called *"debonding"* or *"stripping."* Debonding in hot mixes has been given more attention by researchers than debonding in cold mixes and seal coats, and there has been no documented methodology to quantify bond loss in seal coats. This report contains a description of a newly developed laboratory test experiment called Seal Coat Debonding Test (SDT) to examine the extent of seal coat susceptibility to moisture damage.

When the test was performed on several emulsion based seal-coat specimens, with asphalt emulsion binder, the nature of the outcome strongly suggested dividing seal coats into three categories according to their degree of vulnerability (DV) to moisture damage. Highly vulnerable seal coats are those that experience a loss of aggregate of more than 20% (based on the weight of the base bitumen and the totally intact aggregate). Moderately vulnerable seal coats lose more than 10% and up to 20%, and low vulnerable seal coats lose up to 10%.

The test is relatively easy to conduct and is considered unique in the sense that no similar tests are currently in use to quantify the amount of bond loss in seal coats. The test is applicable only to seal coats using asphalt emulsion as a binder.

KEY WORDS: seal coat, stripping, debonding, laboratory test, moisture damage, emulsion

Despite the different methods used in building asphaltic surfaces, the common goal is to ensure proper bonding between asphalt and aggregate and to maintain a higher retention of the aggregate. The bonding between asphalt and aggregate is a complex phenomenon influenced by the physiochemical and the mechanical properties of both materials. Stripping is also regarded by researchers as a complex mechanism. Several theories have been developed over the years in an attempt to describe the mechanism of bonding and bond failure [*1*].

Assuming that surface water and humidity are inevitable, stripping will continue to be a disturbing problem in the asphalt paving industry. It occurs to hot asphalt mixes as well as cold mixes, in asphalt slabs, and to various types of surface-treatment applications. Stripping of hot asphalt mixes has attracted the attention of several researchers during the last three decades [*2,3*], but unfortunately not enough research has been performed in the area of cold mixes. This report will shed some light on the failure of emulsion based seal coats in terms of bond loss due to moisture damage.

[1] Professor, Department of Civil Engineering, South Dakota State University, Brookings, SD 57007-0495.
[2] Highway engineer, Wisconsin Department of Transportation, 3502 Kinsman Blvd., Madison, WI 53704.

The state of the art in measuring the amount of stripping occurring to asphalt mixes shows that only judgment calls are exercised in placing inferences. Kennedy et al. [3] used a visual inspection technique to estimate the amount of stripped surface in the Texas Boiling test, which is mainly used for hot asphalt mixes. The ASTM Test Method for Coating and Stripping of Bitumen-Aggregate Mixtures (D 1664) employs similar philosophy in measuring stripping. The laboratory technique developed in this research tends to quantify debonding in terms of the amount of aggregate totally separated from the asphalt-aggregate mass that was originally considered in good bond with asphalt before it is subjected to the test environment.

Due to the ever-increasing popularity of emulsions over liquid asphalts in seal coats for environmental and health reasons, this research only employed various types of emulsion in the preparation of the test specimens [4].

Test Procedure

Apparatus

Steel Frame—150 by 150 by 25-mm (6 by 6 by 1-in.) (Fig. 1) to accommodate the 150 by 150-mm (6 by 6-in.) test specimens.

Compaction Head (Fig. 2)—A 150 by 150 by 15-mm (6 by 6 by 0.6-in.) steel plate covered with a 150 by 150 by 13-mm (6 by 6 by 0.5-in.) hard rubber tile. The compaction head is used to deliver the needed pressure to embed and reorient the aggregate particles in the bituminous bed. It is equipped with ball bearings to ensure uniform pressure on the specimen.

FIG. 1—*Steel frame used in making seal-coat specimens:* (A) *shingles,* (B) *steel frame,* (C) *platform.*

FIG. 2—*Compaction head:* (A) *ball bearing,* (B) *compaction head with rubber plate,* (C) *finished specimen.*

Standard Compaction Machine—To apply the necessary compactive effort.

Balance—Capacity of 2 and 5 kg (4.4 and 11 lb) and sensitive to 0.1 g (0.0035 oz).

Constant-Temperature Ovens—Capable of maintaining any temperature between 55 and 149°C (130 and 300°F). Two ovens are used, one to keep the emulsion at 55°C (130°F) and the other to dry the aggregate at 149°C (300°F).

Water Bath—The water bath should be thermostatically controlled so as to maintain the bath at 22 ± 1°C (72 ± 2°F). The bath is used to subject test specimens to the moisture effect by submerging them for a specified amount of time.

Plastic Bags—These should be large enough to house an approximate 150 by 150 by 15-mm (6 by 6 by 0.6-in.) test specimen. The plastic bags should be capable of withstanding a temperature of 60°C (140°F) without distortion and also be the type that can be sealed shut to prevent aggregate particles from escaping without detection. Commercial-size 180 by 230-mm (7 by 9-in.) plastic bags are suitable for this purpose.

Materials

Asphalt Roofing Shingles—Specimens are cut slightly smaller than 150 by 150 mm (6 by 6 in.) to allow easy fit inside the steel frame. A recommended dimension is 149 by 149 mm (5.86 by 5.86 in.).

Aggregate—Prepare the aggregate that meets the applicable specifications for seal coats. In this study an aggregate that 100.0% passes a 9.5-mm (3/8-in.) sieve and is retained on a 6.3-mm (1/4-in.) sieve was used. The aggregate is washed and dried at 149°C (300°F) to constant weight. Store in airtight containers until required for use.

Distilled Water—For use in water bath.

Base-Bitumen—This refers to the emulsion used in making the seal coat. It should be stored at 55°C (130°F) temperature.

Procedure

Assuming that materials are prepared and stored according to the previous outlines, the following steps are executed in the following order:

1. Place the steel frame in the center of a 300 by 300 mm (12 by 12 in.) platform (plywood or any other suitable material).
2. Weigh the asphalt roofing shingle to the nearest 0.1 g (0.0035 oz) and record the weight ($= w_1$), then place it inside the frame so as the rough surface is facing upward.
3. Calculate the desired amount of emulsion to be spread over the test specimen. In this research, 32 g (1.128 oz) of emulsion were used, which is equivalent to 1.36 L/m^2 (0.3 gal/yd^2). Also, determine the desired amount of dry aggregate to be spread over the test specimen. In this study, 371 g (13 oz) of either dry Quartzite or pit-run gravel was used, which is equivalent to 13.6 kg/m^2 (30 lb/yd^2).
4. Place the platform with the steel frame and shingle on a balance and record the total weight. Add the specified amount of emulsion by observing the correct balance reading. Immediately spread the emulsion evenly over the shingle with the aid of a flat head spreading edge, as seen in Fig. 3. Caution should be exercised so as to not pour more bituminous binder than needed. Allow only about two extra grams to compensate for the emulsion that will inevitably stick to the spreading edge.
5. Observe the scale reading after spreading the emulsion. With the aid of a flat, spreading edge, evenly spread the desired amount of dry aggregate over the emulsion as soon as the brownish color of the emulsion turns black.
6. The test specimen should be compacted next. Figure 4 shows the specimen under compaction. A total load of 7.2 kN (0.72 tonf) is applied. This load produces 0.32 MPa (~45 psi) of pressure, which is commonly used in compacting seal coats in the field. Four applications are made to resemble a typical four passes of pneumatic rollers over seal coats.
7. The test specimen is carefully removed from the steel frame with the aid of a sharp knife to separate the specimen edges from the inside walls of the steel frame. The loss of any aggregate or small amounts of emulsion is irrelevant at this stage.
8. The test specimen is then allowed to cure and set at room temperature for at least 24 h (depending on the type of emulsion used). Fans can be used to accelerate the setting process of the emulsion and to develop the total bond between the aggregate and the bitumen.
9. Excess and loose aggregate should be removed by shaking the specimen, tapping it gently on the sides, and rubbing the surface. The rubbing action is very important to the integrity of the test: vigorous and excessive rubbing is meaningless, but too gentle and soft rubbing is not recommended either. A recommended way of rubbing is to hold the specimen flat on one hand and, in a circular motion, rub the surface of the specimen with the other hand. The edges must be trimmed with a sharp knife to ensure that no loose aggregate particles are present. Test specimen weight (shingle plus intact aggregate plus cured bitu-

FIG. 3—*Adding the various components to the shingle:* (A) *spreading edge and* (B) *shingle.*

FIG. 4—*Compacting the specimen.*

minous binder) is then recorded to the nearest 0.1 g (0.0035 oz) (= w_2). The term initial weight, as seen in Table 1, refers to the combined weight of only the remained cured bituminous binder and the totally intact aggregate. This weight is represented by w ($w = w_2 - w_1$).

10. Each test specimen is placed inside a sealable plastic bag. The bag is partially filled with distilled water and seal-locked to prevent the possible escape of any aggregate particles. The plastic bag is then placed in the water bath at 22°C (72°F) for a two-day period (48 h) with the aggregate surface facing downward to guarantee that the aggregate surface is fully submerged. The water bath should be filled with distilled water to a depth of about 100 mm (4 in.), as seen in Fig. 5.

11. The specimen is gently removed from the plastic bag and placed on a platform. If any aggregate particles become loose inside the bag while the specimen is being submerged, they should also be removed from the plastic bag and placed on the platform. Rubbing action is applied again as in Step 9 to remove all loose aggregate particles from the specimen. The separation of aggregate from the specimen at this point is mainly attributed to the loss of bond between the aggregate and bituminous binder. All loose aggregate is carefully collected, oven-dried, and weighed to the nearest 0.1 g (0.0035 oz) (= X_i).

12. Steps 10 and 11 are repeated until only insignificant or no further loss of aggregate is encountered. Individual 48-h losses (X_i) and cumulative loss are recorded (ΣX_i). Laboratory testing of various seal-coat specimens revealed that aggregate loss (X_i) in most specimens will cease or become negligible after ten days (240 h). The total aggregate loss (ΣX_i) should be recorded. The total percent loss is calculated and termed the degree of vulnerability (DV), which is expressed mathematically as DV = $\Sigma X_{i/w}$, where

DV = degree of vulnerability,
ΣX_i = cumulative weight of separated aggregate in grams, and
w = initial weight of intact aggregate and bituminous binder in grams.

TABLE 1—*Summary of results.*

Treatment	Initial Weight,[a] g[b] w	Individual 2-day Period Weight Loss, X_i										Cumulative 10 day Loss, ΣX_i	
		2 days		4 days		6 days		8 days		10 days			
		g	%	g	%	g	%	g	%	g	%	g	%
A-1	147.3	2.9	2.0	2.2	1.5	1.0	0.7	1.0	0.7	1.1	0.8	8.2	5.6
A-2	144.7	2.0	1.4	1.8	12.0	1.4	1.0	0.8	0.6	1.7	1.2	7.7	5.3
B-1	131.9	2.9	2.2	1.0	0.8	1.1	0.8	1.7	1.3	0.3	0.2	7.0	5.3
B-2	121.1	4.6	3.8	1.8	1.5	2.3	1.9	0.8	0.7	0.4	0.3	9.9	8.2
C-1	152.4	11.7	7.7	14.4	9.5	1.5	1.0	1.0	0.7	0.0	0.0	28.6	18.8
C-2	141.2	14.1	10.0	12.3	8.7	1.1	0.8	0.0	0.0	0.0	0.0	27.5	19.5
D-1	148.8	4.9	3.3	7.6	5.1	6.6	4.4	2.2	1.5	0.0	0.0	21.3	14.3
D-2	161.4	8.7	5.4	9.2	5.7	3.3	2.0	0.04	0.3	0.0	0.0	21.6	13.4
E-1	147.1	14.6	9.9	14.0	9.5	3.2	2.2	0.9	0.6	0.0	0.0	32.7	22.2
E-2	142.3	19.1	13.4	13.3	9.4	4.9	3.4	0.8	0.6	0.0	0.0	38.1	26.8
F-1	155.7	12.6	8.1	7.8	5.0	6.0	3.9	4.2	2.7	0.0	0.0	30.6	19.7
F-2	140.6	14.1	10.0	6.2	4.4	4.8	3.4	1.1	0.8	0.0	0.0	26.2	18.6

[a] Totally intact aggregate plus cured bituminous binder.
[b] 1 g = 0.035 oz.

FIG. 5—*Subjecting specimens to moisture effect in a water bath:* (A) *finished specimen inside a sealable bag partially filled with distilled water and* (B) *water bath.*

Application of Test Procedure

For the developed methodology to be examined, several seal coats were constructed. No specific attention was given to the compatibility of the aggregate and the emulsion used in constructing the seal-coat specimens. High float, cationic, and anionic emulsions were used as well as natural aggregate (pit-run gravel), plain quartzite, and hydrated lime-treated quartzite. Table 2 shows the materials used in constructing duplicates of six different seal-coat specimens.

After each of the twelve seal-coat specimens was constructed and subjected to the test, the results were documented as shown in Table 1.

Conclusion

When the test procedure was performed on the duplicate of the six different seal coats, the mean aggregate loss of each treatment, whether after two days interval or at the commencement of the test, varied so widely that establishing various categories of seal coats was warranted. Some seal-coat specimens lost as low as only 5.3% and others as high as 26.8% of the initial weight of the bitumen-aggregate mass. By inspecting all specimens at the conclusion of the test, it was decided that three categories of classification are considered adequate to describe the degree of susceptibility to moisture damage.

The term "degree of vulnerability" is used here to describe how susceptible the emulsion based seal-coat specimens are to moisture damage. Various seal-coat specimens will be categorized as low, medium, or high. Low represents those seal coats that lose aggregate, by debonding, up to 10.0% of their initial weight at the end of 10 days. Medium encompasses

TABLE 2—*Materials used in building seal-coat specimens.*

Test Specimens	Bituminous Product	Aggregate
A-1	cationic emulsion	quartzite
A-2	cationic emulsion	quartzite
B-1	cationic emulsion	quartzite treated with hydrated lime[a]
B-2	cationic emulsion	quartzite treated with hydrated lime[a]
C-1	high float emulsion	quartzite
C-2	high float emulsion	quartzite
D-1	high float emulsion	natural aggregate (pit-run gravel)
D-2	high float emulsion	natural aggregate (pit-run gravel)
E-1	high float emulsion	quartzite treated with REDICOTE 82-S[a]
E-2	high float emulsion	quartzite treated with REDICOTE 82-S[a]
F-1	anionic emulsion	natural aggregate (pit-run gravel)
F-2	anionic emulsion	natural aggregate (pit-run gravel)

[a] To minimize the stripping problem, aggregates were chemically treated with anti-stripping additives. The proper dosage was applied to the aggregate before drying it.

those seal coats that experience a loss of aggregate between 10.1 and 20.0% at the end of 10 days. Highly vulnerable seal coats are those who lose more than 20% of the initial weight in 10 days.

Table 3 shows the classification of the various seal coat specimens in terms of DV, and Fig. 6 is a graphical presentation of the results on a standard chart designed particularly for that purpose. Note that each line represents the average of the two specimens (or more if desired).

An engineering judgment should be exercised to classify those specimens with a DV value in the upper teens because they might be better classified as highly vulnerable, especially when other factors are considered such as climate, traffic volume, percent trucks, etc.

The test method presented in this report is very simple and offers a sound quantitative approach to a problem that was often considered a classical, qualitative problem. It can be

TABLE 3—*Seal-coat classification.*

Treatment	Degree of Vulnerability to Moisture Damage
A-1 A-2	low
B-1 B-2	low
C-1 C-2	medium
D-1 D-2	medium
E-1 E-2	high
F-1 F-2	medium

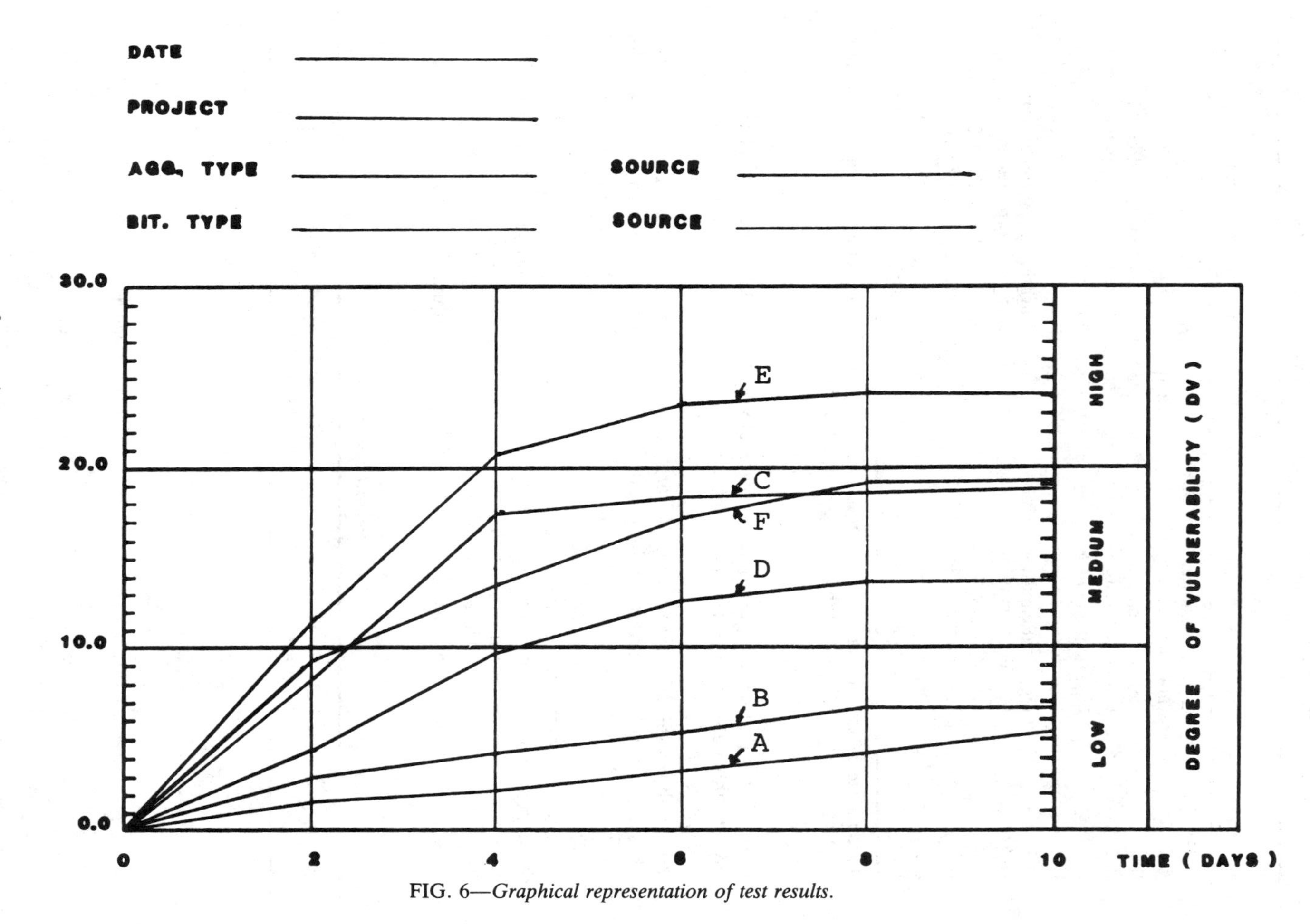

FIG. 6—*Graphical representation of test results.*

used to assist engineers in choosing the right materials emulsion based for seal-coat (chip seal) construction. Further research to correlate laboratory results with actual field performance is currently in progress. The ultimate success of this method is when prediction of field performance can be made through the laboratory test method.

It should be mentioned that the test methodology described in this report is only applicable when an asphalt emulsion is used as the binder. The validity of this method of testing has not been confirmed when other binders are used, and further research work is suggested.

Acknowledgment

The authors would like to thank the Quartzite Rock Association of South Dakota for their financial support of this research.

References

[*1*] Majidzadeh, K. and Frederick, M. "State of the Art: Effect of Water on Bitumen Aggregate Mixtures," Special Report No. 78, Highway Research Board, 1968.

[*2*] Kennedy, T., Roberts, F., and Anagnos, J., "Texas Boiling Test for Evaluating Moisture Susceptibility of Asphalt Mixtures," Research Report 253-5, Research Project 3-9-79-253, Center for Transportation Research, University of Texas at Austin, Jan. 1984.

[*3*] Kennedy, T., Robert, F., Lee, K., and Anognos, J., "Texas Freeze-Thaw Pedestal Test for Evaluating Moisture Susceptibility for Asphalt Mixtures," Research Report 253-3, Research Project 3-9-79-253, Center for Transportation Research, University of Texas at Austin, Feb. 1982.

[*4*] Selim, A., "Debonding and Skidding Characteristics of the Sioux Quartzite in Seal Coats," Final Report, Project #415582, Civil Engineering Department, South Dakota State University, Jan. 1986.

Kuo-Hung Tseng[1] *and Robert L. Lytton*[2]

Prediction of Permanent Deformation in Flexible Pavement Materials

REFERENCE: Tseng, K.-H. and Lytton, R. L., **"Prediction of Permanent Deformation in Flexible Pavement Materials,"** *Implication of Aggregates in the Design, Construction, and Performance of Flexible Pavements, ASTM STP 1016,* H. G. Schreuders and C. R. Marek, Eds., American Society for Testing and Materials, Philadelphia, 1989, pp. 154–172.

ABSTRACT: This paper presents a method to predict the permanent deformation (rutting) in pavements using a mechanistic-empirical model of material characterization. Three permanent deformation parameters are developed through material testing to simply represent the curved relationship between permanent strains and the number of load cycles. Equations are developed by regression analysis which determine how these three parameters are affected by the material properties, environmental conditions (moisture and temperature), and stress state. These relations are important in calculating the permanent deformation of pavement layers since the relation between permanent deformation and cycles of load from the laboratory is usually examined in test conditions that are significantly different from field conditions. The permanent deformations calculated from the method presented are compared with results measured in the field in Florida and are found to be accurate.

The permanent deformation is calculated as the sum of the resilient strains multiplied by the fractional increase of total strains for each material layer of the pavement. The resilient strains of pavement structures under highway loadings are calculated using a finite-element analysis which incorporates both linear and nonlinear stress-strain behavior of the pavement component materials. The fractional increase of total strains is basically in terms of three parameters which characterize the permanent deformation relations from the laboratory. The values of these parameters are developed for pavement materials such as asphalt concrete, gravel and crushed-stone base course materials, and subgrade soils from a variety of data sources. The statistical equations for the three parameters are developed for each type of material represented in the data. The most important terms in the equations are the asphalt content, temperature, resilient modulus, and stress state for asphalt concrete material, and the water content, resilient modulus, and stress state for base and subgrade soils, respectively. These equations and the method as mentioned above has been programmed into a modified ILLI-PAVE program to calculate the resilient strains and permanent deformation in each layer of pavement, taking into account realistic distributions of tire contact pressures, both vertical and horizontal.

The permanent deformation obtained is shown to be in reasonable agreement with the measured results. It is demonstrated that this method provides an appropriate and realistic analysis of prediction of the permanent deformation, and further, the results are used in the prediction of the loss of serviceability index of pavements using the American Association of State Highway and Transportation Officials (AASHTO) Road Test relation. The paper demonstrates the importance of accurate materials characterization in predicting the rutting of asphalt concrete pavements on granular base course.

KEY WORDS: permanent deformation, rut depth, finite-element analysis, resilient strain, asphalt concrete, granular base, subgrade soil, single-axle load, moisture content, asphalt content, ILLI-PAVE computer program

[1] Research assistant, Texas Transportation Institute, Texas A&M University, College Station, TX 77843.

[2] Professor and research engineer, Texas Transportation Institute, Texas A&M University, College Station, TX 77843.

The prediction of permanent deformation in flexible pavements depends upon knowing the physical properties of the layered materials as they are affected by the properties of the aggregates and by varying environmental conditions. In this paper, the prediction is based upon a new method of characterizing permanent deformation in terms of three parameters and upon the use of this new method in a finite-element analysis of the pavement structures. The permanent deformation prediction procedure is included in a modified version of the ILLI-PAVE computer program [*1,2*], which has been revised to include the ability to predict the fatigue cracking and loss of serviceability index of in-service pavements taking into account realistic distributions of tire contact pressure, both vertical and horizontal. The current version of ILLI-PAVE has the ability to use linear and nonlinear characterization of material properties, to use multiple tires and multiple axles, and to provide different amounts of resistance to slip between layers. The program is used in this paper to demonstrate the effect of aggregate characteristics by predicting the permanent deformation of several in-service pavements and comparing the results with the measured values in Florida test road sections.

Permanent Deformation Characterization

The method used to characterize permanent deformation of the pavement materials is in terms of three parameters, that is, ϵ_0, β, and ρ. These parameters are developed by fitting a curve that relates permanent strains to loading cycles obtained from creep and recovery or repeated load triaxial tests. A typical graph of permanent strain versus number of load cycles is shown in Fig. 1. The curve describing this relationship is represented by

$$\epsilon_a = \epsilon_0 e^{-(\rho/N)^{\beta}} \quad (1)$$

Permanent Strain, ϵ_a

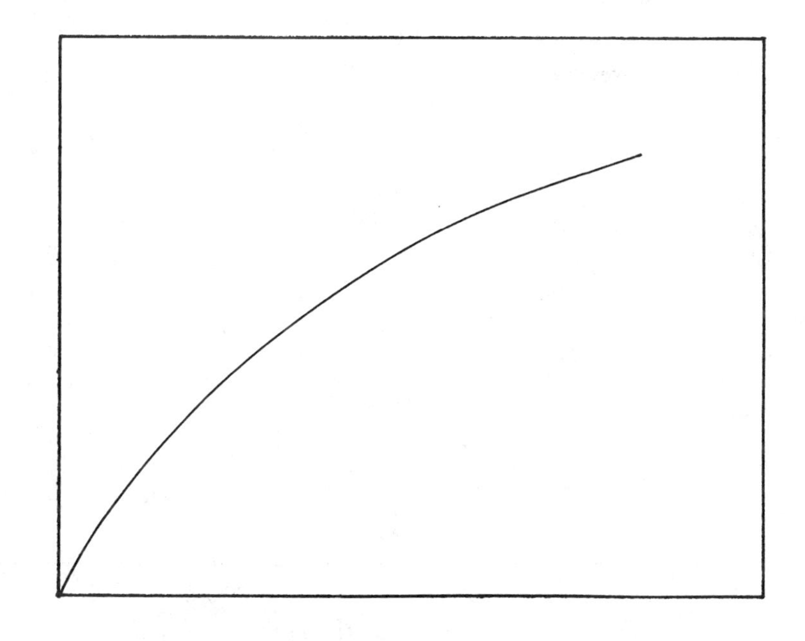

Loading Cycles, N

FIG. 1—*Relationship between permanent strain and loading cycles.*

where

ϵ_a = permanent strain,
N = number of load cycles, and
ϵ_0, β, and ρ = material parameters.

The values of ϵ_0, β, and ρ are different for each sample, depending on the type of the materials and their physical properties, and conditions such as temperature during the testing.

These three parameters can be obtained from the SAS NLIN program [3], which produces least-square estimates of the parameters of a nonlinear model. However, another accurate and simple method to calculate these parameters has been developed. By taking the derivative and manipulating Eq 1, it becomes the equation of a straight line

$$\log[\Delta(\ln \epsilon_a)/\Delta(\ln N)] = \log(\beta\rho^\beta) - \beta \log N \qquad (2)$$

in which β is the slope of line computed by Eq 2. Once β is known, then ρ can be calculated easily from the first term on the right hand side of Eq 2. Once β and ρ are given, ϵ_0 can be computed from Eq 1 by averaging the values of permanent strains against number of load cycles. Studies comparing measured deformations and predicted values are reported elsewhere [2], where they are shown to yield favorable results. Equation 1 has been found to be applicable to all flexible pavement materials: asphalt concrete, granular base, and subgrade soils. Figures 2, 3, and 4 show the results of the use of the three parameter equation.

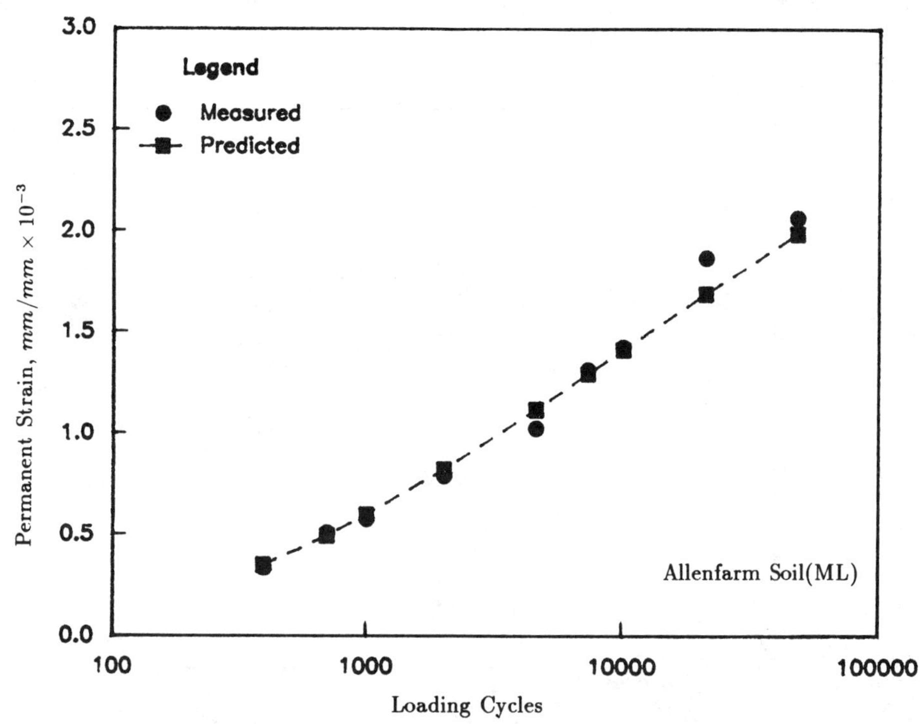

FIG. 2—*Comparisons of predicted and measured permanent strains versus loading cycles for subgrade soil.*

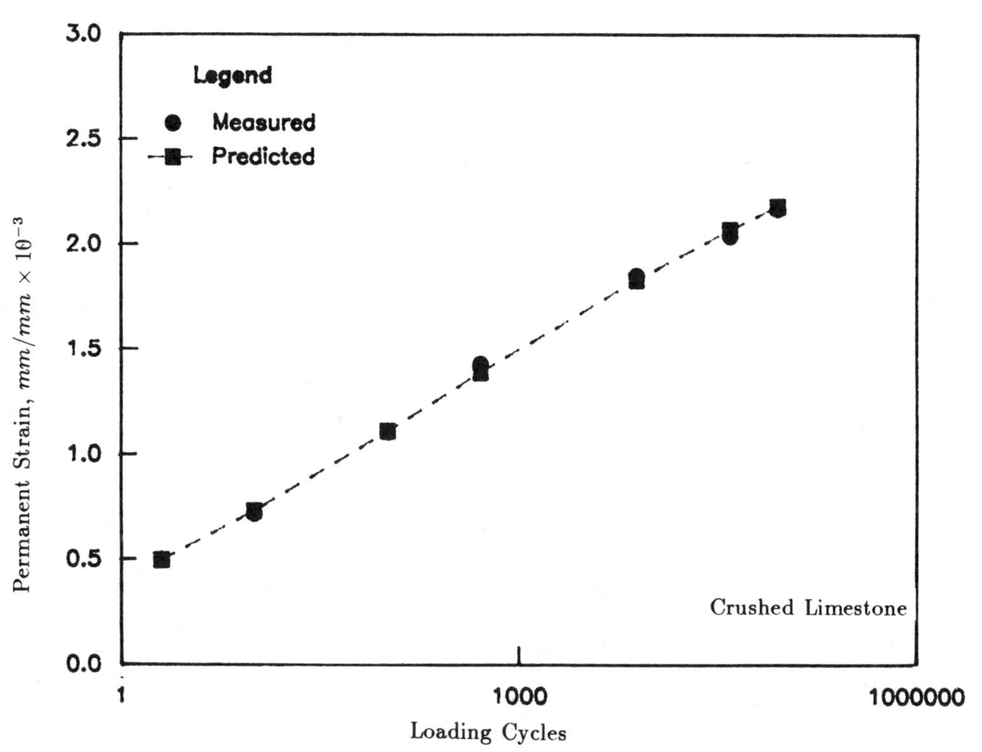

FIG. 3—*Comparisons of predicted and measured permanent strains versus loading cycles for base course materials.*

Permanent Deformation Prediction

Permanent deformation (rut depth) in the wheel path of a flexible pavement is attributed to the accumulation of permanent strains produced by repetitive traffic loads. The model of permanent deformation is based on an evaluation of the vertical resilient strain in each layer by the finite-element method and on the fractional increase of total strains for each material layer of the pavement as determined by the three material properties, ϵ_0, ρ, and β. The finite-element analysis is used to take both linear and nonlinear stress-strain behavior of the materials into account. This approach can be applied to not only a single-axle load, but also multiple axle loads on the surface. The mathematical derivation of the equations to predict permanent deformation for a single axle load as well as multiple axle loads is described elsewhere [*2*].

For a single axle load, the permanent deformation, δ_a, is given by

$$\delta_a(N) = \sum_{i=1}^{n} \left\{ \left(\frac{\epsilon_{0_i}}{\epsilon_{r_i}}\right) e^{-\left(\frac{\rho_i}{N}\right)^{\beta_i}} \int_{d_{i-1}}^{d_i} \epsilon_c(z)\, dz \right\} \tag{3}$$

where

n = number of pavement layers,
ϵ_{ri} = resilient strain imposed in the laboratory test to obtain the three parameters of the material in the ith layer,
N = expected number of load cycles,
d_i = depth of ith layer, and
ϵ_c = vertical resilient strain in the layer i from the finite-element solution.

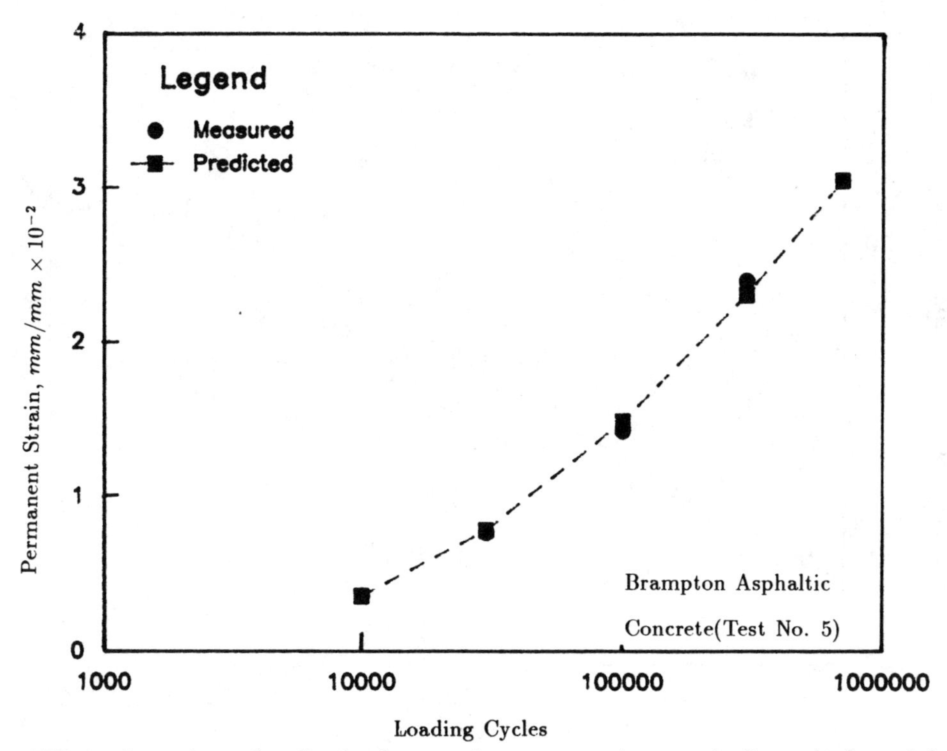

Loading Cycles

FIG. 4—*Comparisons of predicted and measured permanent strains versus loading cycles for asphalt concrete.*

The term

$$\frac{\epsilon_{0_i}}{\epsilon_{r_i}} e^{-\left(\frac{\rho_i}{N}\right)^{\beta_i}}$$

is defined as the fractional increase of total strains. The integral on the right side of Eq 3 can be solved numerically using the trapezoidal rule of integration for the given vertical strain of each element beneath the center of tire loads.

For a tandem axle load with single wheels, the equation of permanent deformation is expressed as

$$\delta_a(N) = \sum_{i=1}^{n} \left\{ \frac{\epsilon_{0_i}}{\epsilon_{r_i}} e^{-\left(\frac{\rho_i}{N}\right)^{\beta_i}} \int_{d_{i-1}}^{d_i} \left(1 + \frac{\Delta\sigma(z)}{\sigma_{max}(z)}\right) \epsilon_c(z)\, dz \right\} \quad (4)$$

The term $\Delta\sigma$ is the difference between σ_{max} and σ_{min}. σ_{max} is determined by superposition as the vertical stress under a single wheel plus the overlap vertical stress at a distance corresponding to the tandem-axle spacing. Since the distribution of the vertical stresses is assumed to be symmetric and the interaction effect of the dual tires is ignored, σ_{min} is simply twice the vertical stresses at half the tandem spacing. The individual values of σ_{max}, σ_{min}, and ϵ_c vary with the depth of the pavement and the size of tire loads. Thus, the estimate for total permanent deformation at the surface can be calculated numerically layer by layer from Eq 4.

Similar to the case of a dual axle load, Eq 4 can be extended for other multiple axle configurations and may be expressed as

$$\delta_a(N) = \sum_{i=1}^{n} \left\{ \frac{\epsilon_{0_i}}{\epsilon_{r_i}} e^{-\left(\frac{\rho_i}{N}\right)^{\beta_i}} \int_{d_{i-1}}^{d_i} \left(1 + \sum_{j=1}^{k-1} \frac{\Delta\sigma_j}{\sigma_{\max}}\right) \epsilon_c(z)\, dz \right\} \tag{5}$$

where k is the number of axles in each axle group and $\Delta\sigma_j$ is the stress difference between the jth and $(j + 1)^{st}$ axle group.

Predictive Equations for ϵ_0/ϵ_r, ρ, and β

As noted previously, the values of ϵ_0/ϵ_r, ρ, and β are material constants derived from a permanent deformation test. To obtain appropriate values of ϵ_0/ϵ_r, ρ, and β for the material of each pavement component, it is necessary to determine how these three parameters are affected by the stress state, density, moisture content, asphalt content, temperature, and other material and environmental characteristics. The effects of these factors are important in calculating permanent deformation of the pavement layers because the laboratory test conditions are significantly different from the actual field conditions. The test conditions affect the relation between the permanent strain and the number of loading cycles obtained from the laboratory, which affects the three parameters and the magnitude of the calculated permanent deformation. In view of this, a comprehensive literature review of permanent deformation test data of pavement materials reported by other researchers has been conducted. From the available data collection, pavement materials are categorized in this study mainly into three groups: asphalt concrete, granular base material, and subgrade soils. A preliminary regression analysis [3] of ϵ_0/ϵ_r, ρ, and β in terms of the available variables is performed for each type of material. Several forms of equations which define relations between the variables are established in the analysis. The most reliable equations relating the three parameters (based on the highest R^2 and lowest standard error) are determined using multiple regression analysis and are discussed below.

Asphalt Concrete

Several permanent deformation data sources for asphalt concrete are available [4–6] and can be reduced to yield values of ϵ_0/ϵ_r, ρ, and β, as shown in Table 1. All of the available data sets are obtained from repeated load triaxial tests except the Florida data set, which is obtained using creep testing. The resilient modulus in Morris's data source is taken from Rauhut's estimation [6]. As can be seen in Table 1, β is generally less than 0.3, and ρ has a range from 10 to 10^{17}. A plot of the relation between β and log ρ for the entire 55 data sets is shown in Fig. 5. A definite correlation appears to exist between increasing ρ and decreasing β for asphalt concrete. It is noted that the Morris data (Canadian Mix) for β in the same range of ρ are much higher than the β values from the other data sets. The significance of this relation may also be used to denote the relative degree of material nonlinearity and to represent extremely variable physical properties. The value of ϵ_0/ϵ_r varies from 10 to 10^3, depending on the material properties of the individual mixes and their stress state.

A preliminary analysis shows that ϵ_0/ϵ_r, ρ, and β are most sensitive to resilient modulus and deviator stress, but the parameters are also sensitive to asphalt content and temperature.

TABLE 1—*Permanent deformation data for asphalt concrete.*

Data Source and Material Type	Density, pcf	A_c, %	T, °F	σ_3, psi	σ_d, psi	σ_θ, psi	E_r ksi	ϵ_r	ϵ_0	ρ	β
Morris,		5.8	90.0	20.0	45.0	105.0	120	3.75×10^{-4}	0.2794	7.916×10^{5}	0.2665
Brampton AC		5.8	65.0	20.0	45.0	105.0	580	7.76×10^{-5}	0.09353	1.146×10^{6}	0.2495
		5.8	65.0	20.0	45.0	105.0	580	7.76×10^{-5}	0.02883	7.905×10^{4}	0.2393
		5.8	90.0	20.0	25.0	85.0	120	2.08×10^{-4}	0.07500	8.106×10^{4}	0.2377
		5.8	77.5	50.0	5.0	155.0	270	1.85×10^{-5}	0.00800	1.130×10^{5}	0.1933
		5.8	77.5	50.0	5.0	155.0	270	1.85×10^{-5}	0.00400	5.131×10^{6}	0.0119
		5.8	77.5	50.0	5.0	155.0	270	1.85×10^{-5}	0.00515	5.233×10^{4}	0.2672
		5.8	77.5	13.6	31.4	72.2	270	1.53×10^{-4}	0.00377	1.688×10^{4}	0.4060
Brown and Snaith,		4.0	68.0	0.0	21.7	21.7	210	1.03×10^{-4}	0.02803	1.175×10^{7}	0.0974
Dense bitumen		4.0	68.0	0.0	58.0	58.0	210	2.79×10^{-4}	0.03668	6.022×10^{5}	0.1299
Macadam		4.0	68.0	29.0	64.0	152.0	210	3.05×10^{-4}	0.03391	9.113×10^{4}	0.1308
		4.0	68.0	14.0	80.0	122.0	210	3.81×10^{-4}	0.05244	5.662×10^{4}	0.1747
Barksdale,	149.9	4.8	95.0	30.0	30.0	120.0	68	4.41×10^{-4}	0.00573	1.054×10^{2}	0.1433
Georgia black	151.6	4.8	89.0	30.0	30.0	120.0	76	3.95×10^{-4}	0.00498	7.605×10^{1}	0.1654
Base, AC-20	150.6	4.8	105.0	30.0	30.0	120.0	100	3.00×10^{-4}	0.00339	2.268×10^{1}	0.2736
	149.3	4.8	85.0	15.0	30.0	75.0	69	4.35×10^{-4}	0.01008	7.155×10^{2}	0.1010
	150.0	4.2	95.0	15.0	30.0	75.0	59	5.08×10^{-4}	0.01102	7.193×10^{3}	0.0838
	150.9	4.8	95.0	15.0	30.0	75.0	100	3.00×10^{-4}	0.00932	2.252×10^{2}	0.0998
	150.1	4.2	105.0	15.0	30.0	75.0	275	1.09×10^{-4}	0.05061	1.444×10^{10}	0.0694
	150.5	4.8	95.0	30.0	30.0	120.0	404	7.43×10^{-5}	0.00751	3.825×10^{2}	0.1419
	150.5	4.8	95.0	5.0	15.0	30.0	227	6.61×10^{-5}	0.00250	9.825×10^{1}	0.1754
	152.0	5.5	95.0	5.0	15.0	30.0	191	7.85×10^{-5}	0.31000	4.906×10^{6}	0.0970
	148.0	4.2	95.0	5.0	15.0	30.0	323	4.64×10^{-5}	0.00149	2.023×10^{1}	0.2362
	149.7	4.2	95.0	5.0	20.0	35.0	242	8.26×10^{-5}	0.00291	5.367×10^{1}	0.2729
	148.4	4.2	95.0	5.0	25.0	40.0	259	9.65×10^{-5}	0.00580	3.564×10^{2}	0.1297
	148.6	4.8	95.0	5.0	25.0	40.0	223	1.12×10^{-4}	0.01060	2.184×10^{2}	0.2431
	151.4	5.5	95.0	5.0	25.0	40.0	220	1.14×10^{-4}	0.00424	2.252×10^{1}	0.2460
	150.3	4.8	95.0	30.0	30.0	120.0	268	1.12×10^{-4}	0.00412	2.266×10^{3}	0.1090
	148.9	4.8	85.0	5.0	25.0	40.0	240	1.04×10^{-4}	0.00707	7.804×10^{3}	0.1216
	151.5	4.2	105.0	5.0	25.0	40.0	20	1.25×10^{-4}	0.02369	9.637×10^{6}	0.0778
	150.3	4.8	95.0	5.0	32.0	47.0	232	1.38×10^{-4}	0.00700	1.831×10^{2}	0.1569
	150.4	4.8	95.0	5.0	37.0	52.0	258	1.43×10^{-4}	0.01416	2.178×10^{4}	0.1758
	150.7	4.2	95.0	5.0	25.0	40.0	253	9.88×10^{-5}	0.00583	1.964×10^{4}	0.1174
	150.6	4.8	95.0	5.0	25.0	40.0	91	2.75×10^{-4}	0.00516	1.755×10^{1}	0.1838
Rauhut, AC-20	151.5	4.7	100.0	10.0	20.0	50.0	1 250	1.60×10^{-5}	0.02570	3.524×10^{12}	0.0674
	151.3	4.4	100.0	10.0	20.0	50.0	1 880	1.06×10^{-5}	0.00881	3.032×10^{7}	0.1075
AC-20	145.2	4.8	100.0	10.0	20.0	50.0	629	3.18×10^{-5}	0.03810	4.138×10^{12}	0.0598
AC-5	145.6	4.5	100.0	10.0	20.0	50.0	1 070	1.87×10^{-5}	0.06920	9.318×10^{12}	0.0668
AC-30	153.9	5.3	100.0	10.0	20.0	50.0	756	2.65×10^{-5}	0.10280	5.626×10^{11}	0.0728
AC-20	150.4	8.2	100.0	10.0	20.0	50.0	310	6.45×10^{-5}	0.12460	6.827×10^{12}	0.0686
AC-10	136.1	4.8	70.0	10.0	20.0	50.0	260	7.69×10^{-5}	0.00454	6.599×10^{2}	0.1303
AC-10	151.1	6.4	100.0	10.0	20.0	50.0	175	1.14×10^{-4}	0.00997	2.763×10^{3}	0.1927
Florida,	140.3	6.5	41.0	0.0	35.0	35.0	2 187	1.60×10^{-5}	0.00284	1.455×10^{15}	0.0663
Type I AC	141.7	6.5	77.0	0.0	10.0	10.0	625	1.60×10^{-5}	0.00787	1.027×10^{16}	0.0496
	141.7	6.5	104.0	0.0	2.0	2.0	163	1.25×10^{-5}	0.00303	3.368×10^{16}	0.0476
Sand asphalt	125.7	7.5	41.0	0.0	10.0	10.0	813	1.23×10^{-5}	0.00802	1.562×10^{15}	0.0670
Hot mix	128.4	7.5	77.0	0.0	2.5	2.5	208	1.20×10^{-5}	0.00081	4.397×10^{5}	0.1325
	121.7	8.0	77.0	0.0	17.5	17.5	227	7.70×10^{-5}	0.02287	3.144×10^{13}	0.0686
	121.2	8.0	41.0	0.0	35.0	35.0	722	4.85×10^{-5}	0.00908	1.338×10^{15}	0.0648
	122.9	8.0	41.0	0.0	10.0	10.0	962	1.04×10^{-5}	0.01611	1.381×10^{17}	0.0588
	122.5	8.0	77.0	0.0	2.5	2.5	194	1.29×10^{-5}	0.02201	3.527×10^{16}	0.0587
	124.1	8.5	41.0	0.0	10.0	10.0	952	1.05×10^{-5}	0.00055	5.893×10^{6}	0.1272
	126.5	8.5	104.0	0.0	1.0	1.0	35.7	2.80×10^{-5}	0.00479	1.932×10^{16}	0.0451
	121.9	8.0	104.0	0.0	1.0	1.0	53.2	1.88×10^{-5}	0.01096	3.833×10^{16}	0.0542
	127.9	7.5	41.0	0.0	10.0	10.0	909	1.10×10^{-5}	0.00317	6.914×10^{14}	0.0666

NOTE: 1 psi = 6.9 kPa.

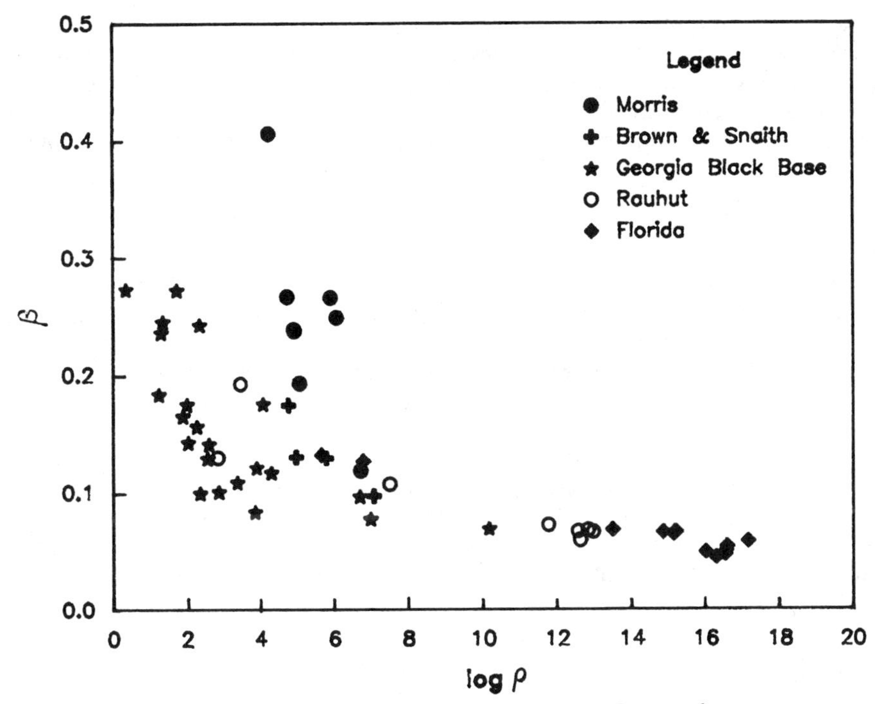

FIG. 5—*Relation between log ρ and β in AC materials.*

Because of this, several forms of equations including the more sensitive variables are considered in the multiple regression analysis of each parameter in terms of asphalt content, temperature, deviator stress, and resilient modulus. The most accurate equations of these three parameters are

$$\log\left(\frac{\epsilon_0}{\epsilon_r}\right) = -5.04349 + 0.01812A_c + 0.011045A_c^2 + 0.01127T - 0.203249 \log \sigma_d + 1.12228 \log E_r \quad (6)$$
$$R^2 = 0.44$$

$$\log \rho = 8.105675 - 4.241965A_c + 0.54159A_c^2 + 0.03865T - 0.014874\sigma_d + 0.000005E_r \quad (7)$$
$$R^2 = 0.62$$

$$\log \beta = -2.51475 + 0.60816A_c - 0.05282A_c^2 - 0.00214T + 0.16597 \log \sigma_d - 0.0000002E_r \quad (8)$$
$$R^2 = 0.43$$

where

A_c = asphalt content, %,
T = temperature, °F,
σ_d = deviator stress, psi, and
E_r = resilient modulus, psi.

Although the equations above are not exceptionally accurate from a statistical viewpoint, they are indicative of the relationship between the variables and give some estimates of the effects of these variables. It is apparent that other variables related to the type, shape, and surface texture of the aggregate must be included in order to derive more accurate equations. Unfortunately, these data are not customarily measured or recorded in the repeated load testing programs reported in the literature. The equations show, as expected, that permanent deformation increases with increasing temperature, deviator stress, and asphalt content.

Granular Base Material

Table 2 reflects available permanent deformation data from different sources [*6–9*] for granular materials that can be reduced to yield the values for ϵ_0, ρ, and β. All experimental data in these data sets come from repeated load triaxial tests. As can be seen, the value of ρ is generally around 10^3, and β is less than 0.3. ϵ_0/ϵ_r is generally less than 60, which implies that permanent deformations of granular materials under the same strain level may be less than that of asphalt concrete or subgrade soils. Statistical analysis indicates that ρ and β are independent of each other.

An analysis of the entire 16 data sets shows that deviator stress, bulk stress, moisture content, and resilient modulus have the most significant effects on ϵ_0/ϵ_r, and β. The parameter, ρ, is not sensitive to any of the factors. Multiple regressions are run with ϵ_0/ϵ_r, β, and ρ, respectively, since the dependent variable and bulk stress, moisture content, and resilient modulus as independent variables. The most accurate equations of these three parameters are

$$\log\left(\frac{\epsilon_0}{\epsilon_r}\right) = 0.80978 - 0.06626W_c + 0.003077\sigma_\theta + 0.000003E_r \quad R^2 = 0.60 \tag{9}$$

$$\log \beta = -0.9190 + 0.03105W_c + 0.001806\sigma_\theta - 0.0000015E_r \quad R^2 = 0.74 \tag{10}$$

$$\log \rho = -1.78667 + 1.45062W_c - 0.0003784\sigma_\theta^2 - 0.002074W_c^2\sigma_\theta - 0.0000105E_r \quad R^2 = 0.66 \tag{11}$$

TABLE 2—*Permanent deformation data for granular base material.*

Data Source and Material Type	Density, pcf	W_c, %	σ_3, psi	σ_d, psi	σ_θ, psi	E_r, ksi	ϵ_r	ϵ_0	ρ	β
Barksdale,	137.0	4.2	10.0	46.0	76.0	37.5	0.001230	0.01688	3.375×10^3	0.1756
Crushed prophyrite	137.0	4.2	10.0	28.3	58.3	32.6	0.000868	0.00510	2.242×10^2	0.2319
Granite	137.0	4.2	10.0	19.4	49.4	29.8	0.000651	0.00398	1.779×10^3	0.1661
	137.0	4.2	10.0	15.0	45.0	28.4	0.000528	0.00329	8.870×10^3	0.1592
Chisolm and Townsend	133.9	2.4	25.0	116.0	191.0	189	0.000614	0.02710	6.093×10^3	0.1200
Crushed limestone	140.9	2.4	10.0	45.9	75.9	120	0.000383	0.00849	31.04	0.1370
	142.6	2.4	20.0	41.4	101.4	167	0.000248	0.00335	1.996×10^2	0.1400
Gravelly sand	131.7	4.5	10.0	46.4	76.4	109	0.000426	0.01076	1.638×10^3	0.1300
	132.8	5.6	10.0	32.6	62.6	90	0.000362	0.01150	3.493×10^2	0.1250
Kalcheff and Hicks	144.0	5.0	5.0	15.0	30.0	46	0.000326	0.00212	2.853×10^3	0.1904
Pa. 2A subbase	144.0	5.0	5.0	15.0	30.0	45	0.000333	0.00043	6.596×10^3	0.1628
	144.0	5.0	20.0	15.0	30.0	48	0.000313	0.00113	3.856×10^3	0.1835
	144.0	5.0	20.0	60.0	120.0	116	0.000517	0.00633	2.255×10^3	0.1992
	144.0	5.0	20.0	60.0	120.0	114	0.000526	0.00414	2.382×10^3	0.1977
Florida,	117.2	10.0	10.0	20.0	50.0	37	0.000541	0.00138	1.052×10^3	0.2858
Limerock	116.6	10.0	10.0	20.0	50.0	37	0.000541	0.00122	7.303×10^2	0.2759

NOTE: 1 psi = 6.9 kPa.

where W_c = water content, %, and σ_θ = bulk stress, psi. It is noted that a factor for deviator stress is not included in the equations, although the parameters are sensitive to deviator stress. However, in the multiple regression analysis, inclusion of deviator stress as a variable produces a higher standard error than other variables. The equations for granular material imply that permanent deformation increases with increasing water content and increasing bulk stress. This corroborates the conclusions made by Rauhut et al. [*10*].

Subgrade Soil

A total of 26 individual permanent deformation data sets for subgrade soil are found in the literature from three different sources [*10–12*] that can be reduced to yield values of ϵ_0, ρ, and β. All permanent deformation data are from repeated load triaxial tests (Table 3). Since resilient modulus (resilient strain) is not included in the Monismith data sets, a total of 20 data sets is considered in this study. The value of β from these data sets is generally less than 0.2, while ρ has a range from 10^2 to 10^{26}. Figure 6 shows the relation between β and log ρ and indicates that a definite correlation exists between increasing ρ and decreasing β. The relation may imply that the soils, classified by the Unified Soil Classification as CH, CL, and ML, can be categorized into groups associated with the three-parameter equation (Eq 1).

The preliminary analysis of the R^2 of the regression equations indicates that water content and deviator stress are the most significant factors which affect the values of ρ and β, whereas ϵ_0/ϵ_r is also sensitive to resilient modulus. Further studies of the interaction effects of those

TABLE 3—*Permanent deformation data for subgrade soils.*

Data Source and Material Type	Density, pcf	W_c, %	σ_3, psi	σ_d, psi	σ_θ, psi	E_r, ksi	ϵ_r	ϵ_0	ρ	β
Rauhut, compacted clay	107.8	11.4	2.5	1.5	9.0	35	4.286×10^{-5}	0.3953	9.915×10^{11}	0.1084
Silty clay	121.4	15.2	2.5	1.5	9.0	22.7	6.608×10^{-5}	0.0751	8.773×10^{9}	0.1091
Edris and Lytton,	79.6	25.2	15.0	10.0	55.0	17.4	0.000575	0.8706	1.841×10^{12}	0.1050
Clay - CH	80.1	25.7	3.5	13.7	24.2	12.9	0.00106	1.683	1.337×10^{13}	0.0930
	79.2	24.7	20.0	15.0	75.0	6.87	0.00218	0.1443	2.884×10^{12}	0.0863
	83.4	35.3	15.0	10.0	55.0	8.2	0.00122	11.756	7.730×10^{17}	0.0690
	83.6	35.3	3.5	13.7	24.2	9.07	0.00151	1.237	3.826×10^{18}	0.05660
	83.0	35.7	20.0	15.0	75.0	10	0.00150	6.514	1.827×10^{26}	0.04145
Clayey silt - ML	112.6	13.7	15.0	10.0	55.0	8.12	0.00123	0.1024	2.938×10^{9}	0.1111
	113.1	13.5	3.5	13.7	24.2	10	0.00137	0.0228	5.801×10^{5}	0.1854
	113.0	13.7	20.0	15.0	75.0	7.05	0.00213	0.0927	1.715×10^{9}	0.1094
	101.9	10.0	15.0	10.0	55.0	9.5	0.00105	0.0206	2.625×10^{6}	0.1911
	110.6	12.6	3.5	13.7	24.2	8.64	0.00159	0.8940	3.190×10^{13}	0.0860
	106.9	10.8	20.0	15.0	75.0	7.61	0.00197	0.0213	1.118×10^{7}	0.1575
Silty clay - CL	96.9	6.6	3.5	13.7	24.2	11.8	0.00116	0.0105	5.477×10^{5}	0.1700
	100.2	13.6	15.0	10.0	55.0	17.2	0.00058	0.0414	9.496×10^{9}	0.1034
	105.3	17.2	3.5	13.7	24.2	15.6	0.000878	0.4512	4.541×10^{6}	0.1784
	106.2	19.8	3.5	13.7	24.2	11.2	0.00123	0.1567	2.425×10^{8}	0.1351
	99.5	8.4	20.0	15.0	75.0	13	0.00115	0.0037	8.952×10^{3}	0.2755
	85.7	30.0	20.0	15.0	75.0	11.5	0.00131	3.398	7.549×10^{15}	0.0764
Monismith, silty clay	107.0	16.4	5.0	5.0	20.0			0.0079	7.762×10^{11}	0.0793
	107.0	16.5	5.0	10.0	25.0			0.0105	4.859×10^{18}	0.0396
	107.0	16.1	5.0	20.0	35.0			0.00532	8.381×10^{8}	0.0612
	107.0	19.3	5.0	5.0	20.0			0.00617	1.519×10^{11}	0.0556
	107.0	19.7	5.0	10.0	25.0			0.41620	4.510×10^{25}	0.0341
	107.0	19.3	5.0	20.0	35.0			0.03413	4.748×10^{2}	0.20250

NOTE: 1 psi = 6.9 kPa.

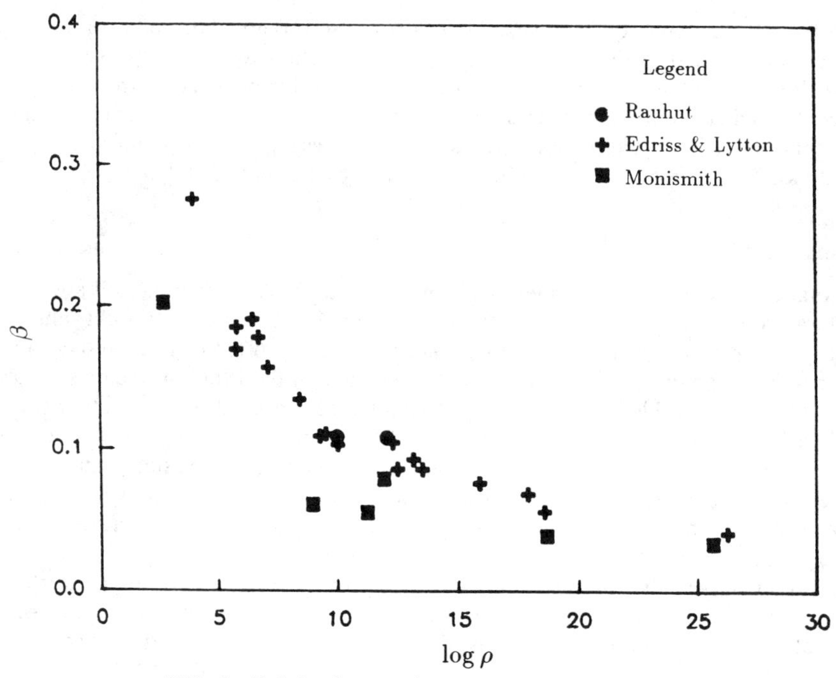

FIG. 6—*Relation between log ρ and β for subgrade soils.*

variables on the parameters show that increased accuracy can be attained for ρ and β, but not for ϵ_0/ϵ_r. Based on the above studies, the most accurate equations for ϵ_0/ϵ_r, ρ, and β obtained by using multiple regression are

$$\log\left(\frac{\epsilon_0}{\epsilon_r}\right) = -1.69867 + 0.09121 W_c - 0.11921\sigma_d + 0.91219 \log E_r \qquad R^2 = 0.81 \tag{12}$$

$$\log \rho = 11.009 + 0.000681 W_c^2\sigma_d - 0.40260\sigma_d + 0.0000545 W_c^2\sigma_\theta \qquad R^2 = 0.86 \tag{13}$$

$$\log \beta = 0.9730 - 0.0000278 W_c^2\sigma_d + 0.017165\sigma_d - 0.0000338 W_c^2\sigma_\theta \qquad R^2 = 0.74 \tag{14}$$

These equations show that permanent deformation greatly increases with increasing water content and deviator stress. In the modified ILLI-PAVE computer program, water content is input dependent on seasonal variations. This takes into account the fact that water content during the spring thaw condition is quite different from that in the dry-hot condition.

Application and Results

The Lake Wales test road in Florida was selected to demonstrate the use of the modified ILLI-PAVE computer program in calculating permanent deformation based on the model described previously. Construction of the Lake Wales test road (a four-lane facility) was completed in January 1971. The total number of 80-kN (18-kip) axle loads was approximately 0.27 million in the northbound traffic lane in 1971. The input of traffic data in terms of

80-kN (18-kip) equivalent single axle loads (EASL)/day is about 728, 771, 816, and 886, respectively, for the first four years. All computer program analyses are based on a 40-kN (9000-lb) uniform circular load as a representation of the two dual wheels of a standard 80-kN (18-kip) single-axle load.

The various base and surface thickness combinations are included in the test sections. The limerock base thicknesses are 10.2, 15.2, 20.3, and 25.4 cm (4, 6, 8, and 10 in.), and asphalt concrete thicknesses are 3.8 and 7.6 cm (1.5 and 3 in.). The relationship between resilient moduli of the surface layer and temperatures expected to be encountered at the Lake Wales test road is considered in selecting modulus values. Average monthly temperatures were obtained from U.S. Weather Bureau records, and each year was divided into two seasons, summer (April to October) and winter (November to March). The characteristics of the limerock base are stress-dependent and are described as a function of the bulk stresses. Although the characteristics of the subgrade soil are not correlated to deviator stress from laboratory data, resilient moduli are dependent on the bulk stress. The major input of the material properties for each layer is summarized as shown in Table 4.

Figure 7 shows comparisons of permanent deformation (rut depth) between measured values and those predicted with modified ILLI-PAVE. As can be seen, the modified ILLI-PAVE permanent deformation prediction is excellent for all test sections.

Effects of Aggregate Characteristics in Permanent Deformation

The regression equations of the three parameters as described above are derived from the entire sets of data for each layer of pavement materials. The effects on permanent deformation of the different aggregate types can also be demonstrated. To do this, it is necessary to shift the general regression equations (Eq 6 to 11) of the three parameters to a particular material with a specific type of aggregate. The "constant" term in each regression equation for each material is obtained by substituting the observed values of the variables (Tables 1 and 2) into the equations and calculating the "constant." Tables 5 and 6 show the "constant" term of each parameter for the selected aggregate types, asphalt concrete and granular base material, respectively.

The modified ILLI-PAVE program is used to predict what the permanent deformation of the Lake Wales test road (Section No. 5B) would be if layer materials with different aggregate types had been used. Two hypothetical cases are analyzed, the first using asphalt concrete with different aggregate types in the surface layer which lies on the limerock base and second using a granular base course with different types of aggregate underlying the standard Florida Mix asphaltic concrete surface layer. The thickness of asphalt concrete and base course are 7.6 and 20.3 cm (3 and 8 in.), respectively, corresponding to Test Section No. 5B. The properties of subgrade soil, environmental conditions, and traffic loads used in the analyses are the same as used in calculating the rut depths in Fig. 7. Figure 8 shows the results of using different base materials in the test road. As can be seen, if the gravelly

TABLE 4—*Input data of resilient moduli of each pavement material, and temperature variation.*

Season Variation	Lengh of Month	Average Temperature, °F (°C)	Resilient Moduli, MPa		
			Surface	Base	Subgrade
Summer	7	78.0 (25.5)	4134	$E_r = k_1 \sigma_\theta k_2$	
Winter	5	63.2 (17.3)	7235	$k_1 = 143.0$	$k_1 = 98.0$
				$k_2 = 0.335$	$k_2 = 0.358$

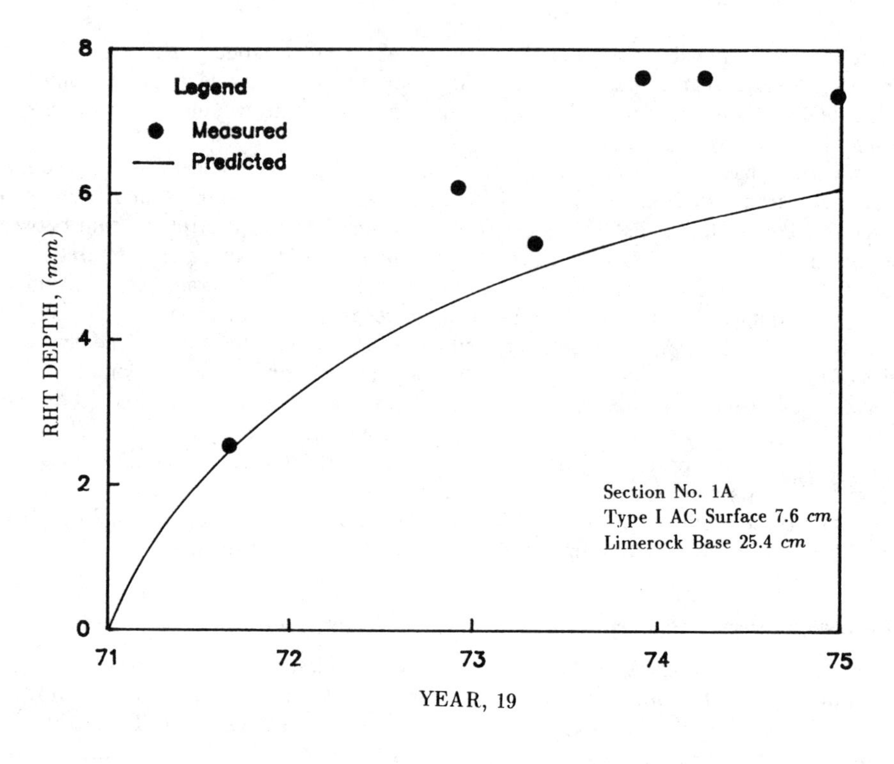

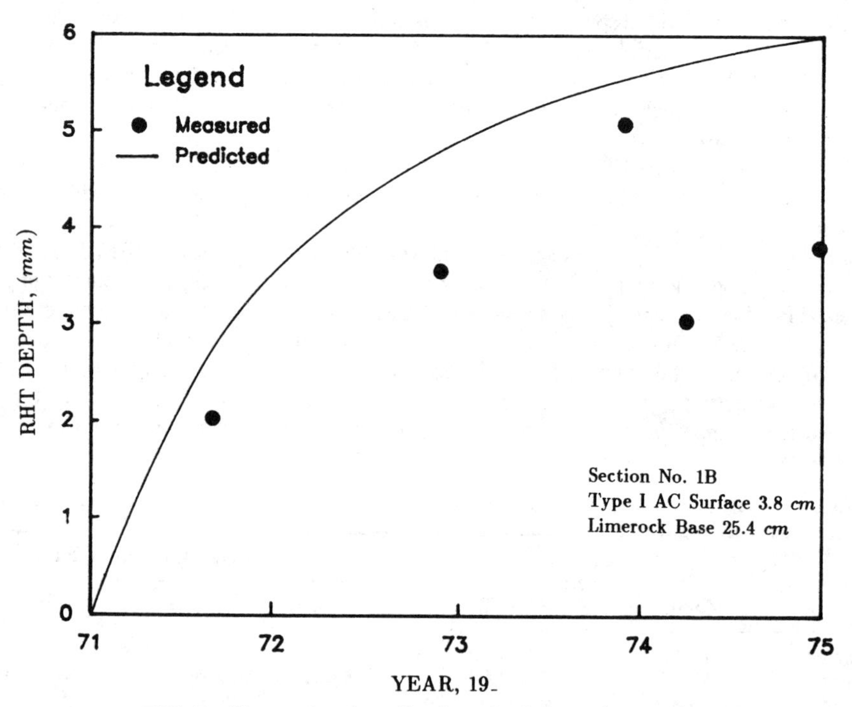

FIG. 7—*Measured and predicted rut depth for various sections.*

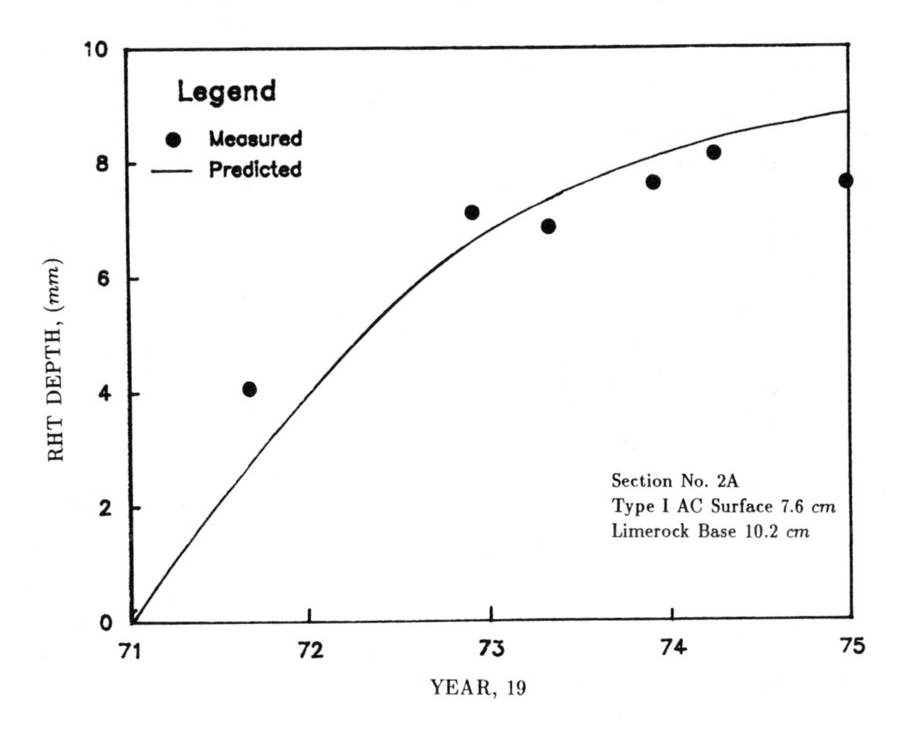

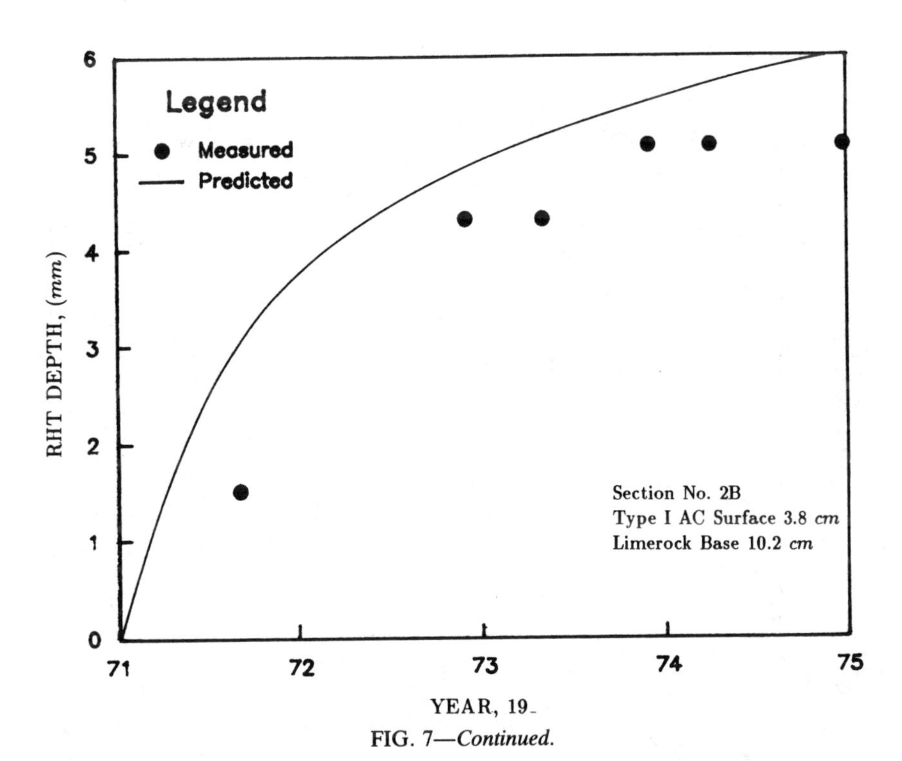

FIG. 7—*Continued.*

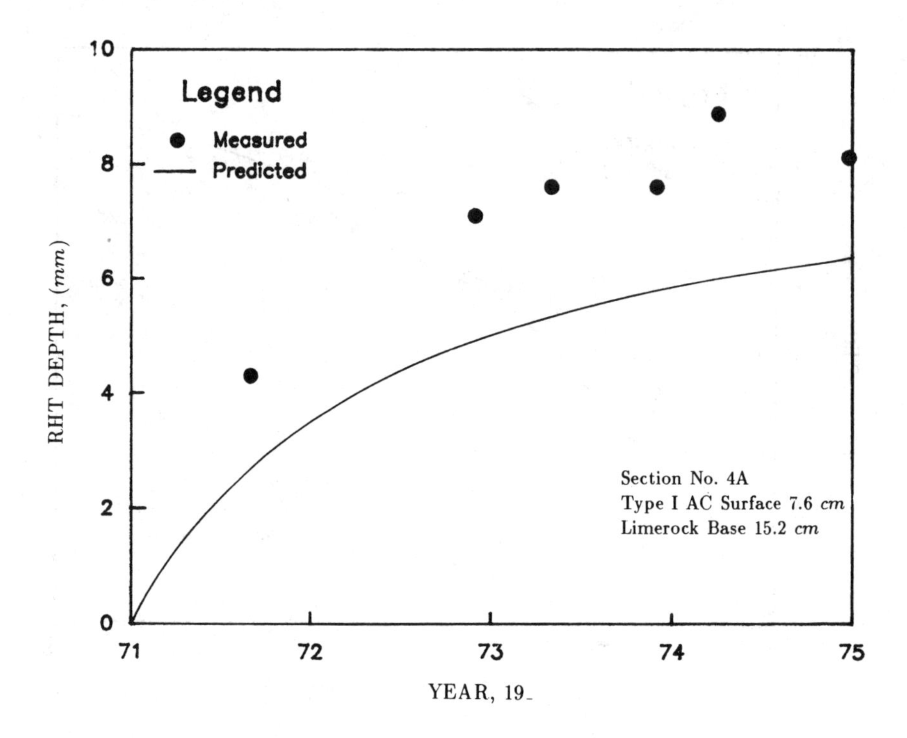

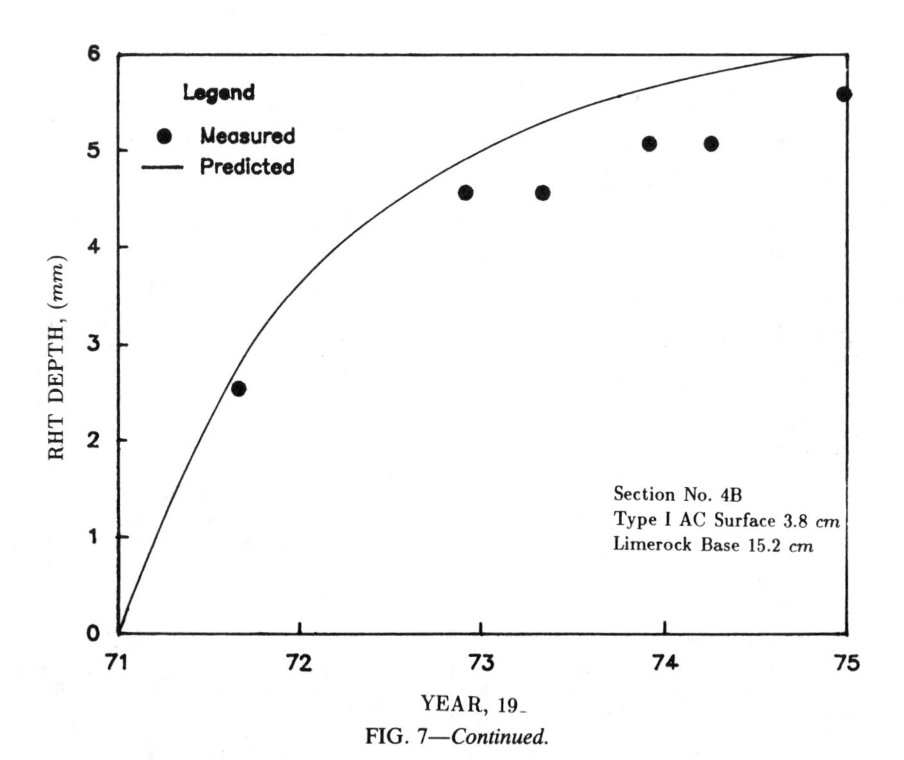

FIG. 7—*Continued.*

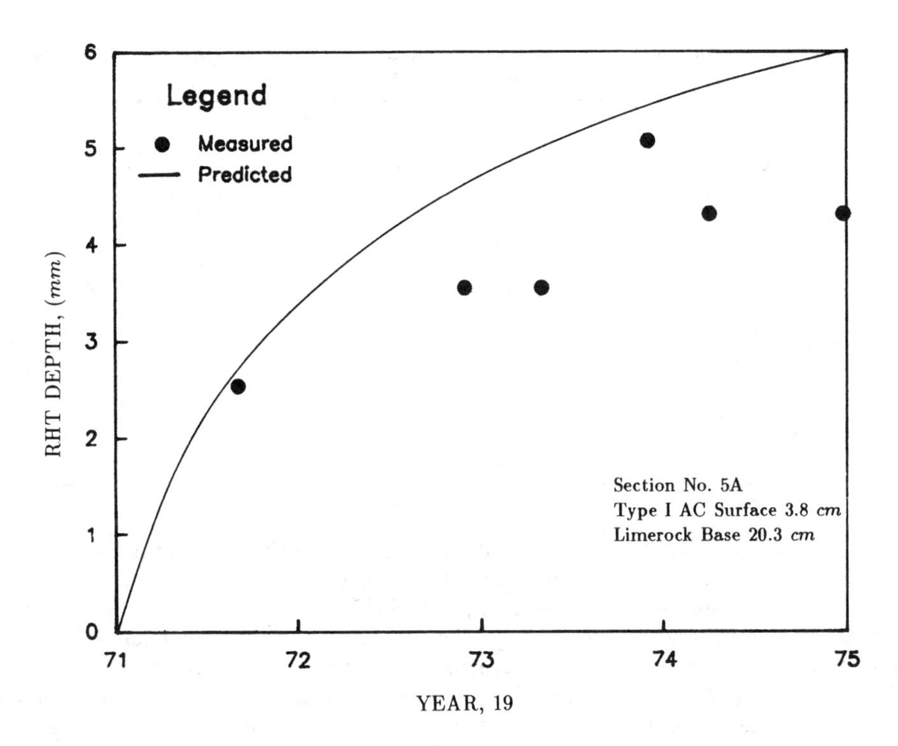

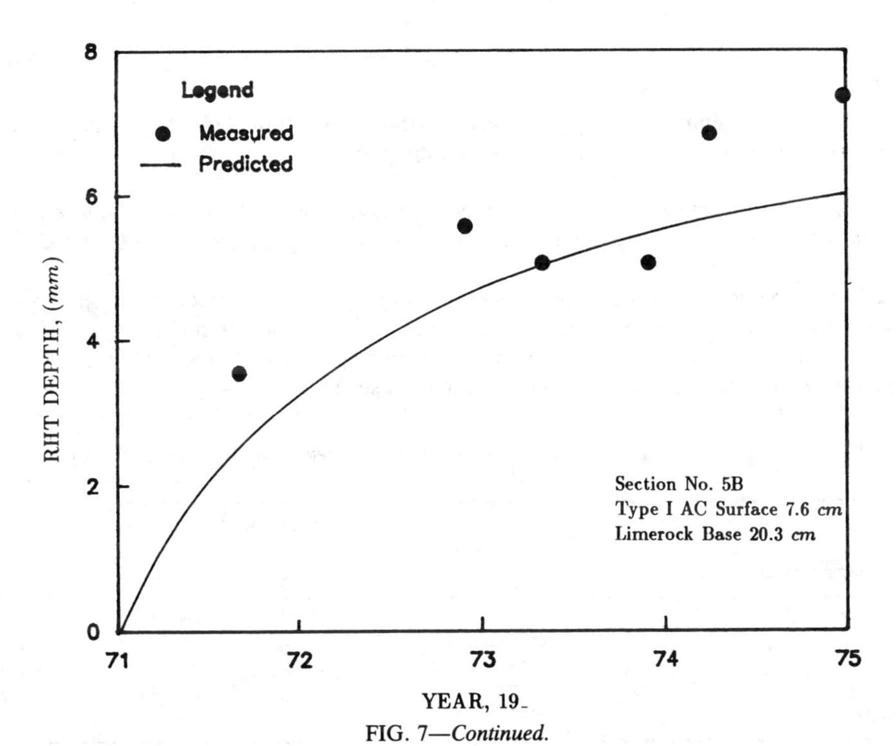

FIG. 7—*Continued.*

TABLE 5—*Constants in regression equations of the three parameters for asphalt concrete with different aggregates.*

Aggregate Type	Constant		
	ϵ_0/ϵ_r	ρ	β
Brampton Asphalt Concrete	−4.6927	7.5319	−2.3572
Dense bitumen Macadam	−4.4492	11.0960	−2.5673
Georgia Balck base	−5.2871	6.8364	−2.5144
Dense graded asphalt Concrete, AC-20	−4.8280	10.7030	−2.6150
Sand asphalt Hot mix	−5.0727	8.3112	−2.5316
OVERALL	−5.0435	8.1057	−2.5148

sand is used as base course in the test road, the permanent deformation at the surface will be higher than that of using any other types of aggregate. On the other hand, using sand-asphalt hot mix as base course, as is done frequently in Florida, a lesser amount of permanent deformation compared with other aggregates will occur after 1.2 million equivalent 80-kN (18 kip) axle loads. Figure 9 shows the permanent deformation prediction in the test section using different asphalt concrete mixes. In this illustration, the Brampton asphalt concrete ruts more than the dense bitumen macadam, which in turn ruts more than the dense graded asphalt concrete with the AC-20 binder.

Conclusions

A mechanistic-empirical model for predicting permanent deformation of the flexible pavement materials is presented. In the course of the study, the following conclusions are reached:

1. The three-parameter equation is a good representation of the relation between permanent strains and loading cycles in laboratory tests and is applicable to flexible pavement materials.
2. The regression equations for ϵ_0/ϵ_r, ρ, and β in terms of environmental conditions and stress state are developed and found to be accurate in calculating permanent deformation.
3. The model of permanent deformation predictions presented in the study is compared with values measured at the Lake Wales test road sections. It has been found from this

TABLE 6—*Constants in regression equations of the three parameters for granular base materials with different aggregates.*

Aggregate Type	Constant		
	ϵ_0/ϵ_r	ρ	β
Crushed Prophyrite granite	0.8782	−1.6852	−0.9275
Crushed limestone	0.7559	−1.8427	−0.8924
Gravelly sand	1.2745	−1.6078	−1.0276
OVERALL	0.8098	−1.7867	−0.9190

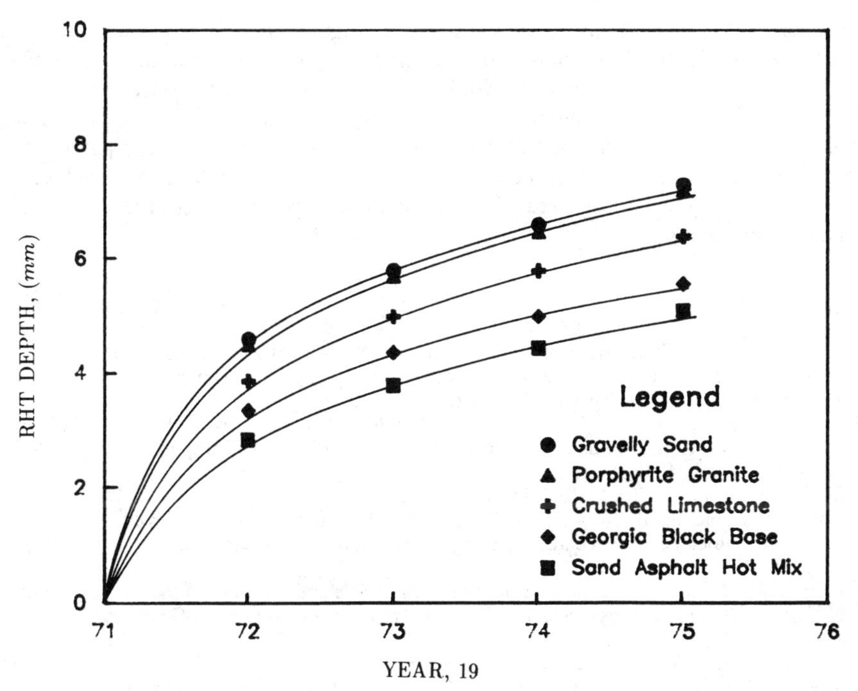

FIG. 8—*Rut depth in Test Section 5B using different base material.*

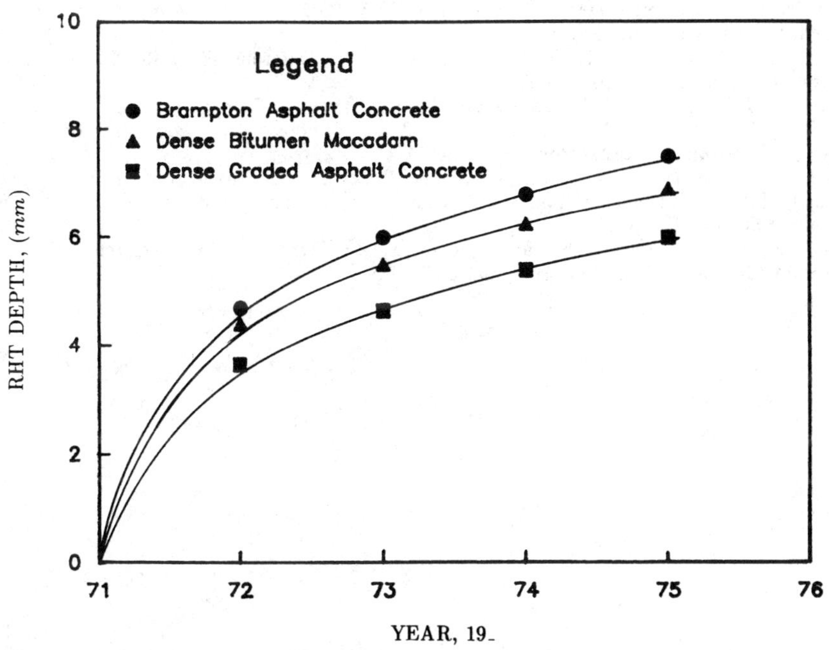

FIG. 9—*Rut depth in Test Section 5B using different asphalt concrete mix.*

comparison that the predicted values for a total of eight test sections are within ±30% of the measured rut depth. This comparison was made with the rut depths at the end of 1975. Rutting in two sections was predicted within −10% of the measured rut depth.

4. Hypothetical analyses using materials whose properties have been published show, not surprisingly, that varying amounts of rutting may be expected if different aggregates are used in the surface course or base course of a pavement. The results of the analyses show that it is possible by use of the predictions to rank the suitability of aggregates for resisting permanent deformation.

References

[*1*] "ILLI-PAVE—A Finite Element Program for the Analysis of Pavements," Construction Engineering Laboratory and the Transportation Facilities group, Department of Civil Engineering, University of Illinois at Urbana, May 1982.

[*2*] Roberts, F. L., Tielking, J. T., Middleton, D., Lytton, R. L., and Tseng, K.-H., "Effects of Tire Pressures on Flexible Pavements," Research Report 372-1F, Texas Transportation Institute, Texas A&M University, College Station, TX, Aug. 1986.

[*3*] *SAS User's Guide,* 1979 ed., SAS Institute, Inc., Cary, NC.

[*4*] Brown, S. F. and Snaith, M. S., "The Permanent Deformation Characteristics of a Dense Bitumen Macadam Subjected to Repeated Loading," *Proceedings,* Association of Asphalt Paving Technologists, Vol. 43, 1974.

[*5*] Leitner, E. B. and Page, G. C., "Bituminous Materials and Pavement Evaluation," Study 81-4, NCHRP 1-10B, Florida Dynamic Test Results, prepared for Woodward-Clyde Consultants, Dec. 1981.

[*6*] Rauhut, J. B., O'Quin, J. C., and Hudson, W. R., "Sensitivity Analysis of FHWA Structural Model VESYS II," Report No. FHWA-RD-76-23, Vols. 1 and 2, Federal Highway Administration, Washington, DC, March 1976.

[*7*] Barksdale, R. D. in *Proceedings,* 3rd International Conference on Structural Design of Asphalt Pavements, London, 1972, pp. 161–174.

[*8*] Chisolm, E. E. and Townsend, F. C., "Behavioral Characteristics of Gravelly Sand and Crushed Limestone for Pavement Design," Report No. FAA-RD-75-177, U.S. Army Waterways Experiment Station, Vicksburg, MS, Sept. 1976.

[*9*] Kalcheff, I. V. and Hicks, R. G., "A Test Procedure for Determining the Resilient Properties of Granular Materials," *Journal of Testing and Evaluation,* Vol. I, No. 6, 1973, pp. 472–479.

[*10*] Rauhut, J. B., Lytton, R. L., and Darter, M. I., "Pavement Damage Functions for Cost Allocation," Report No. FHWA/RD-84/019, Vol. 2, Descriptions of Detailed Studies, Federal Highway Administration, Washington, DC, June 1984.

[*11*] Edris, E. V., Jr. and Lytton, R. L., "Dynamic Properties of Subgrade Soils, Including Environmental Effects," Technical Report 164-3, Texas Transportation Institute, Texas A&M University, College Station, TX, May 1976.

[*12*] Monismith, C. L., Ogawa, N., and Freeme, C. R., "Transportation Research Record 537," National Research Council, Washington, DC, 1975, pp. 1–17.

Author Index

Subject Index

Q

R

S

T